THROUGH
VEGETAL
BEING

CRITICAL LIFE STUDIES

CRITICAL LIFE STUDIES

JAMI WEINSTEIN, CLAIRE COLEBROOK, AND MYRA J. HIRD,
SERIES EDITORS

The core concept of critical life studies strikes at the heart of the dilemma that contemporary critical theory has been circling around: namely, the negotiation of the human, its residues, a priori configurations, the persistence of humanism in structures of thought, and the figure of life as a constitutive focus for ethical, political, ontological, and epistemological questions. Despite attempts to move quickly through humanism (and organicism) to more adequate theoretical concepts, such haste has impeded the analysis of how the humanist concept life itself is preconfigured or immanent to the supposedly new conceptual leap. The Critical Life Studies series thus aims to destabilize critical theory's central figure life—no longer should we rely upon it as the horizon of all constitutive meaning, but instead begin with life as the problematic of critical theory and its reconceptualization as the condition of possibility for thought. By reframing the notion of life critically—outside the orbit and primacy of the human and subversive to its organic forms—the series aims to foster a more expansive, less parochial engagement with critical theory.

THROUGH
VEGETAL
BEING

Two Philosophical Perspectives

LUCE IRIGARAY
AND
MICHAEL MARDER

Columbia University Press   *New York*

Columbia University Press
Publishers Since 1893
New York   Chichester, West Sussex
cup.columbia.edu

Copyright © 2016 Columbia University Press
All rights reserved

Library of Congress Cataloging-in-Publication Data
Names: Irigaray, Luce, author. | Marder, Michael, 1980–
Title: Through vegetal being : two philosophical perspectives /
Luce Irigaray and Michael Marder.
Description: New York : Columbia University Press, 2006. |
Series: Critical life studies | Includes bibliographical references and index.
Identifiers: LCCN 2016000987 | ISBN 9780231173865 (cloth : alk. paper) |
ISBN 9780231173872 (pbk. : alk. paper) | ISBN 9780231541510 (e-book)
Subjects: LCSH: Plants (Philosophy) | Philosophy of nature. |
Irigaray, Luce, author. | Marder, Michael, 1980–
Classification: LCC B105.P535 I75 2016 | DDC 113—dc23
LC record available at http://lccn.loc.gov/2016000987

Cover design: Evan Gaffney
Cover image: © Jessica Hines, *Spirit Stories # 28*

References to websites (URLs) were accurate at the time of writing.
Neither the author nor Columbia University Press is responsible for URLs that
may have expired or changed since the manuscript was prepared.

CONTENTS

MICHAEL MARDER

PREFACE

I T is our concern about the current state of nature and the living that compelled us to write a book together. If originally we imagined that it would develop from dialogues corresponding to the theme of each chapter, we quickly understood that this plan was too ambitious or not yet suitable for various reasons. Our approaches to the problem were quite different, and to distinguish our positions at theoretical, ethical, and political levels, while treating our common objective, proved to be impossible, especially given that we hardly knew and live far from one another. The dilemma was either to give up our project or to invent another presentation for the volume that suggested development toward a future dialogue. Our proposal, then, was to write a book with the text by Luce and that by Michael presented in an upside-down format, with the prospect of a meeting in the middle of the volume. Unfortunately, this suggestive solution, which inspired our way of rendering possible a dialogue while both being faithful to our perspectives, did not meet with the agreement of the publisher, especially for technical reasons. Thus it will be up to the readers to find the most fecund manner of broaching our contributions, that is, to perceive at once the main message concerning vegetal being and our different ways of dealing with it.

Without wishing to go into what will appear in our specific approaches to each theme, we can already stress some traits that differentiate our respective positions. No doubt, the two of us are grateful for the help that the vegetal world brought to our lives and continues providing us with. It is in regard to how to care for it that our analyses and proposals differ from one another. Michael thinks of the vegetal world as such and searches for the tracks of its presence in the authors of our tradition in order to provide this tradition with a new understanding and impact, and he endeavors to rebase human thinking by taking plant being into consideration. Luce, for her part, focuses more on the need to modify our conception of subjectivity in order to become able to give birth to a new way of being and existing, especially with regard to the whole living world. If the former works on the interpretation of our past philosophy by taking into consideration plant being, the importance of which has been generally neglected, the latter asks for a more radical refoundation of our culture starting from the determination, especially the sexuate determination, of our subjectivity, which allows respect for life and its development. If Michael hopes that a return to the Greek *phusis* would give rise to a new growth for humans, a growth immanent and close to that of plants, including at the level of sexuation, Luce is concerned that such a return be accompanied by the cultivation of human subjectivity, which cannot happen without us taking on a cut, a void, an insuperable negative and ensuring another relation to transcendence with respect to our natural environment and belonging. Whereas Michael lingers on harms and dangers that our current economy imposes on ecology, Luce reminds us of teachings of some Eastern traditions and the laws that Antigone defended as elements valid for an ecological economy that we have to take into account and develop.

These are only few aspects among those that characterize our respective positions. Blurring them through a hasty and approximative dialogue might harm the perception of our thoughts and render it impossible for readers to clear their path toward a new way of being and behaving, which the current state of our environment, of all living beings, and of our human becoming urgently requires.

Our initial proposal for the organization of the book also had the advantage of avoiding the choice between a first and a second author in the display of the volume. In the absence of such a presentation, Luce will appear as the first contributor and Michael as the second. Beyond the fact that this follows the alphabetic order of our surnames, it corresponds to the way in which the writing of the book happened most often, as the letter introducing the volume and many of the following chapters indicate.

August 2015

THROUGH
VEGETAL
BEING

LUCE IRIGARAY

PROLOGUE

Dear Michael Marder,

I came to know you through your invitation to take part in the volume *Deconstructing Zionism* that you coedited with Gianni Vattimo.[1] It is on this occasion that you offered to send to me your *Plant-Thinking*,[2] a book devoted to a meditative reflection on the vegetal world, notably as a path to overcoming our traditional metaphysics without falling into a lack or a relaxation of thinking, as is too often the case.

I lived for such a long time with the vegetal world that, opening your book, I was both surprised and felt at home, even before having read very much of the text. My surprise grew when I received the chapter on my thought that you wrote for *The Philosopher's Plant*,[3] your book on the relations between certain philosophers and flowers. Too often, the comments on my work confine themselves to a kind of appropriation and/or to criticism. While reading your text, I was, if even only partly, sent back to myself and not removed from myself, a thing that deeply touched me. This experience was not at all narcissistic, but, instead, a feeling that may occur in a first encounter. I thought that it presaged the possibility of a relationship between us and its fecundity, perhaps shareable with others through a common work.

One of my failings being too quickly giving way to enthusiasm, I began to wonder about some questions, to which your manner of behaving has already provided me with first answers. One of these questions concerns your position with respect to overcoming nihilism, given what I noted regarding your belonging to the philosophical trend that works on the deconstruction of our metaphysical tradition. I am extremely vigilant on the passage from a culture that was nihilistic, through its subjection of our global being to suprasensitive values, to an epoch presumed to be postmetaphysical. I am afraid that what can then happen might be still more perilous for humanity. All the so-called overcoming or surpassing through a falling back into a mere sensitiveness and its presupposed artistic expression are examples of such a danger. However, I do not believe any more that substituting plurality for oneness is susceptible to realizing such a transition: the multiple and the one often partake in the same logic, and if the one, or the One, is removed from the constitution of thought and subjectivity, it may reemerge, masked, under the guise of a dictatorial leader, even assumed to be a democratic one.

It is true that the way toward giving up nihilism, which Nietzsche rightly condemns, is really difficult to open and to follow. I, for one, think that the cultural elaboration and the ethical practice of the relations between differently sexuated subjects, beginning with two, can act as both a passage and a basic structure for overcoming nihilism, while respecting Nietzsche's teaching. Is it not what Nietzsche himself sensed when he said that he needed a woman to pursue his work? This affirmation has not been taken seriously enough, probably because sexuate difference is, precisely, the unthought of Western culture for which suprasensitive values that Nietzsche critiques have been substituted. Returning to it in order to think about it with the seriousness that it deserves corresponds to devoting ourselves to a cultivation of life. This is still lacking, and we urgently need it today in order to place ourselves anew in relation to the source of our human energy, as well as to construct a culture that would be, possibly, universal without submission to more or less particular suprasensitive requirements. Indeed, sexuate difference presupposes a transcendental level—one that remains sensible because of the involvement of sexuation, which goes beyond a mere physical materiality—through the respect for the irreducible difference between two subjects who do not partake in the same sexuate identity.

You found a limit to deconstruction in the vegetal world, a choice that is both clever and wise, given the danger that life itself runs today. It is what attracted my attention to your work and is at the origin of my proposal regarding a common work. The vegetal world is, from my infancy, what allowed me to survive, but also to find the roots of life again after the expulsion from my social environment that followed the publication of *Speculum*.[4] Taking refuge in the vegetal world also revealed to me some aspects of the beginning of Greek culture that we have forgotten, something that encouraged me to pursue a criticism of our past Western culture, not starting from the neuter or neutralized, and presumably universal, identity that this tradition artificially assigned to me, but from life itself and its necessary sexuation.

No doubt, *Speculum* corresponds to a critique of Western tradition that I was able to carry out thanks to my feminine identity and the "step back"—as Heidegger probably would say—it permitted me to take as a result of my absence from its active weaving. Such expulsion granted me a perspective that is more difficult to reach for the one who got deliberately involved in the construction of our culture. And this could explain why it was necessary for you to return to the plant world, and not simply to yourself as a man, in order to leave Western metaphysics in a manner that is not nihilistic. In my opinion, this choice is more relevant than the very fashionable return to the animal world, because not only does it concern a world that is able to ensure our survival but it also has many things to teach us, as you rightly stress. It is not by chance that some of the greatest spiritual figures have been compared to elements of the vegetal world rather than to animals, at least insofar as the expression of their message. Your focus on the vegetal world also ought to remind you of the fact that you are a living being, thus sexuate, something to which our tradition has barely awakened us, both at the level of life and that of spirituality. Nevertheless, my first question will be: Could your worthy and crucial focus on the plant world be only the reverse of our nihilistic tradition and so form one body with it? But, no doubt, your option is the less perilous one and it preserves the possibility of a new epoch of the becoming of humanity and of its cultivation.

I wonder about another question: How can we speak of the vegetal world? Is not one of its teachings to show without saying, or to say without words? I imagine that we will try to display and signify on this side or beyond any discourse. And this will not be an easy undertaking in a book. This will force us to give up the tradition of a language of philosophy. Will we be capable of such a gesture, such a challenge? I attempt to risk it in my own texts, a thing that brings me a lack of understanding, indeed, the removal from some academic or editorial circles that, instead of realizing that we must henceforth adopt another way of thinking, another logic, desperately cling to information, representation, and science's language that they think they master, whereas it has dominated them for a long time. How will we make the way together in search of another saying, allowing us to pass on a meaning different from the one, now in great part exhausted, into which our tradition initiated us? How to reach together the multiple meanings of words that "burst forth as flowers," as Hölderlin writes in his poem "Brot und Wein"[5]? Such a multiplicity of meanings—for which the publisher of *Speculum* criticized me as though I claimed to write the Bible—bears witness to an intimacy that tries to express without completely revealing, for such revealing would destroy it.

And this leads me to the third question about which I wonder. I think that both you and I have lived and continue to experience intimacy with the vegetal world. You used the word in one of your letters, and it is a key word in many of my texts too, especially in *Sharing the World*.[6] You also wrote, almost as an echo of the wish that I expressed in my first proposal regarding this book, that you would like to pass on to others what you experience of the vegetal world. I do not believe that this experience can be a mere intellectual one, as relevant as it may be. I imagine that it is rather an experience essential to your survival and your becoming. Anyway, it is what it is for me. Hence, the fear, which mingles with desire, of transmitting it. Such a coexistence with vegetal being—I could almost say: this vegetal existence—keeps me alive and secretly goes with my words. Can I communicate it, and how, without betraying it, without forgetting it and forgetting myself in such a forgetting: without losing both it and myself? In other words: Can I still return among humans, and through what path?

Dear Michael Marder, I am really sorry to ask you these questions when I suggested writing this book to you. But our project will be workable, if it is, only by facing the questions that will arise between us. May the intimacy that we both experience with the vegetal world allow us to tackle them with measure, serenity, and fecundity.

Hopefully,

Luce Irigaray
Paris, 24 November–4 December 2013

1

SEEKING REFUGE IN THE VEGETAL WORLD

From my early infancy, the vegetal world has been my favorite dwelling. Undoubtedly, this is the result of the context of my life at that time, but also of some events and of the way humans behaved toward me.

I was born in a little village of coal mines on the border between Belgium and France. My family occupied a company accommodation, provided by the mine where my father worked. My father was half-Belgian and half-Italian, and my mother was two-thirds Belgian and one-third French. They had in common a love of nature and a mining belonging: my father had studied engineering, and he began to work in the mine, the director of which was my mother's father. Neither of them was very demonstrative, but their love of nature and their admiration for my maternal grandfather were obvious. This latter was, also for me, the hero of the family. He was the son of a miner and he himself became a miner at a really young age, after being expelled from high school for his impertinence. While working in the mine, he secretly prepared the entrance examination for the School of Mining Engineering, and he ended up as the director of a mine.

As I was the third child, I was affectionately adopted by this grandfather, the first two children being the favorites of my mother and my father, respectively. My grandfather, too, loved nature very much, and he often took me along for fishing, visiting the nests of birds, or walking in the garden. It is thanks to him that I could undertake my research work: before dying, he asked my parents to promise to accept my desire to study at the university.

I thus spent my childhood in a context in which nature was really valued. Nature was also our main source of recreational activity. As children, we did not have many toys, and we did not wish for more. On and on, we played in nature with what it provided us with. The only fabricated toy that I received brought me more pain than joy: my godmother gave me a magnificent doll, but it was too big, too cold, too rigid—I may as well say too dead—for me, and this gift plunged me into utter despair, which I tried to soothe by playing with animals, above all with rabbits. I behaved with them as with a doll or a baby: I dressed them; I carried them in a small doll carriage; I bottle-fed them, etc. All that was a little demanding, but I was overjoyed with their movements, with their hearts beating, their jumping out of the carriage, their rebellion against my care. They were living! I also devoted much time to butterflies, to ladybugs, to birds, to lambs, and other animals that populated the garden. Perhaps I rediscovered something of a heaven on earth when I preferred life to death. Moreover, it was in the garden that the religious feasts were celebrated: one fir of the garden was used as a Christmas tree, and we had to find the Easter eggs that were hidden in various plants or bushes.

Unfortunately, I was removed from this Garden of Eden by two events: the beginning of my schooling and the preparation for my first communion. These two things made me ill. I fell into a deep tiredness and continuous anxiety, especially into frequent crises of scruples. The little joyful girl of the garden has become a terrified child, withdrawn into herself and with various sensory problems. Finally, my mother got worried about my condition and took me to a specialist who suggested suppressing my homework and sending me into the garden as soon as I returned from school. It was an excellent prescription that, alas! was not accompanied by banning me from going to confession. Hence I recovered only part of my health, my innocence, and my enjoyment from dwelling in the Garden of Eden.

Two other cataclysms happened in my childhood or adolescence when my parents decided to put me in a boarding school or in a summer camp. Each time this ended with being on a hunger strike, something that corresponded to an instinctual gesture more than a conscious decision. Obviously, everyone imagined that my attitude was the result of the absence of my mother, and also, I, myself, believed that it was the case. However, I am not sure that this was the true reason.

The third time the thing happened, when they finally took me back home, my mother was really angry because I had thwarted the plans of the family. I was weak because I had not eaten for some days, but, instead of stretching out her arms and embracing me, she shouted at me: "Go away into the garden!" Amazingly, her behavior did not affect me too much. In the garden, I felt at home and pacified, and I did not ask for more to recover my health. I was just worrying about the possibility that my parents could send me to the boarding school again.

Once more, a pediatrician was my savior, though it was not the case with all the doctors, and later I had to learn how to cure myself. This wise and loving doctor said to my mother, "The girl suffers from a lack of freedom; she needs to attend a day school and not stay locked up in a boarding school." Luckily, for my mother, this prescription became a gospel truth. My everyday life was very hard, because I had to get up very early, to go to the station, and to travel by train for one hour, then walk for one more hour, to finally reach school; and I had to repeat the same journey to return home in the late afternoon. But I was happy with this style of life, which brought me back to my favorite dwelling each evening.

I would like to add, to be historically rigorous, that the event which led to my last hunger strike was the announcement, by letter, at the time when I was still staying at the boarding school, of the death of Moses, a little rabbit that I had saved after it fell into a great basin full of water. Probably I had the feeling that something went wrong with life when I was far away at the boarding school and I had to go back to the garden to care for life itself.

I could give other examples of my need to seek refuge in the vegetal world and ask for assistance from it. Just a little, but significant, event. One day my mother unfairly and heavily punished me for the howls of my little sister. I was still very young, but I left the family home, and even the garden in which the thing happened, and took refuge in the woods, which were a few miles away from there. My mother sent someone to search for me, and it was not so easy to find me and bring me back home. Already at that time, it was not from humans but from nature that I asked for help.

I also think that my decision, during my first university years, to opt for defending my feminine identity and values against the neutralization of the individual, of discourse, and of thought took part in the same longing for living, instead of subjecting myself to cultural constructions that paralyzed life, its growing and its sharing. All that did not follow from a completely aware decision, but from a will to live, the conditions of which I instinctively sensed, whereas my culture ignored and even scorned them, preferring constructed values, which had become societal customs, over the care and cultivation of life.

Even when I was writing *Speculum*, though my choices were firm and corresponded to a living and ethical embodiment, their foundation was not yet totally clear for me. But who can imagine, for example, that I wrote *Speculum* while sitting in the woods or near the sea, and not in a library? However, this remained a vital necessity, more than a requirement for my thought, at least on an absolutely conscious level. This book had to cause my expulsion from the university, from the school of psychoanalysis in which I had my training, and from the various scientific circles I frequented and where I had my friends, so that I began to wonder

about the truth that I was seeking and upholding. Perhaps, in Heideggerian terms, I could say that a bend or a turn—*ein Kehre*—was necessary, so
that I glimpsed something of the horizon that my thinking was opening
or reopening.

    5–10 December 2013

2

A CULTURE FORGETFUL OF LIFE

What happened with *Speculum*? Contrary to my expectations, this book was a great success. I composed it during four years that corresponded to crossing a wilderness. At that time I was still practicing psychoanalysis, and I was also a political activist of women's liberation. Thus I could not devote all my time to writing the book. Furthermore, it was not a book that could be written in a few weeks or months: it called for many readings or rereadings and lengthy meditation. Given the difficulty of the task that the achievement of *Speculum* had represented, I imagined that it would have a very limited readership. I thought that it would, perhaps, have four readers. Why four? I am not sure that *Speculum* got four readers—I have not met them—but it found a crowd of buyers. For example, Les Éditions de Minuit publishing house was forced to open on the Saturday after its publication in order to provide the bookshops with copies that were in short supply.

However, at the same time, I was dismissed from my teaching post at the University of Paris, Vincennes; I was expelled from the Lacanian school of psychoanalysis, to which I belonged, and from the circle of my acquaintances and even of my friends. None of the people from my past relationships celebrated the success of *Speculum* with me. I received many letters from women I did not know, who thanked me for a book that changed their lives, but they signed only with a first name without any address, and I could not be in touch with them. They have very much helped me to endure what happened, as did the fact that I had already been in absolute distress due to a love pain. I had thus a capacity for resisting social rejections that were less crucial for my life—a resistance that some intellectuals lacked, especially some disciples of Lacan who committed suicide. Nevertheless, it was a terrible hardship to undergo! I lost my social reputation, career, and environment; my friends and the psychoanalysts, with whom I had been in training in order to become a psychoanalyst myself, disowned me.

Furthermore, these people held me responsible for all that. No one put the stress on the objectivity of the cultural event that *Speculum* represented, except, in a way, the women who wrote to me to give thanks, without commenting on the cultural relevance of what was happening. Most people of the establishment, especially of the academic establishment, blamed me for my manner of writing or behaving, which, according to them, was the very reason for the criticism and the exclusion I was suffering—and this still remains the case today, with, sometimes, suggestions of possible invention on my part regarding what occurred after the publication of *Speculum*. However, my way of writing or behaving had not changed from one day to the next. What made the difference was the truth that *Speculum* unveiled.

I must admit that I myself did not completely realize the objective significance of this event. Perhaps I could call it, today, a sort of historical or epochal event, that of the advent of two different subjectivities, with their respective objectivities. If I struggled, and continue struggling, to maintain open a new historical perspective, I did not then understand the whole meaning of what was happening.

And I first had to struggle for my survival! It was not a small matter, and, once more, the macrocosm welcomed me to save the microcosm that I was. It was the case with the sky and the sun that became my most constant company, with the vegetal world that shared the air with me and showed me the endurance of life, and also with some animals, more often the wild than the domesticated ones, as I recount in the text "Animal Compassion."[1] In reality, animals sense better than most humans when someone needs help, and they are able to provide assistance, even of a spiritual nature, something that certain cultures know and tell us. It is also said in our tradition, regarding hermits, that they received their food from animals, and the Holy Spirit is represented as a dove. Now, in some Eastern traditions, the candidate for spiritual life receives a teaching from the birds, as I comment in *To Be Two*.[2]

In reality, nature has preserved me against a double danger: dying of dereliction and solitude or vanishing into a mediatic representation. As I write in the text "Between Myth and History: The Tragedy of Antigone": "Fortunately, if I have been excluded from society—from universities, psychoanalytical institutions, circles of scientists and even friends, in part from publishing houses and, more recently, from my house itself— I have not been deprived of my relation to the natural world. Expelled from public organizations, enclosed or shrouded with a silence that I sometimes felt to be the opaque wall of a tomb, I have not been deprived of my relation to the air, to the sun, to the plant and animal worlds. I have been expelled from the polis, the city, the human society to which I belonged and sent back to the natural world that my contemporaries no longer appreciate or consider of much value, and hence something of which it was unnecessary to deprive me. Being sent to the natural world in this way has allowed me to survive or, better, to rediscover what life itself is."[3]

Through the event that *Speculum* provoked, I have been sent back to the thing that my tradition neglected: life itself. The problem was that I had no means, no language, to represent life. Life escapes representation as it escapes the tool of a representational way of thinking: predicative logic. Life has nothing to do with all that, but this "all that" was precisely what my tradition took into account. Expelled by the presumed key players in the institutional and societal worlds, I was also expelled from my cultural background and remained without words for expressing what happened and serving me as a guide.

It is an experience that can be communicated with difficulty. And, sometimes, I thought that being imprisoned would have been less terrible than such being left with or in nothing—I could say: buried in nothingness—if it were not for the will to live, which I badly felt. At that time, it was above all thanks to my faithfulness to a past experience of a positive absolute, to the persons who accompanied and helped me, and to the acknowledgment of my mother's work in giving me birth that I wanted to live. I think that I wanted to live, but I was no longer able to perceive this desire clearly. Nevertheless, I did not want any power, solely to live.

I thus tried to regain life itself, of which my culture and social environment deprived me. The matter was really that of conquering, at any moment, breathing and energy enough for my survival. I mostly received these from the macrocosm: from the air, the sun, the vegetal world. I paid extreme attention to this source of life and searched for a new possible path to relate to, or with, it. I felt a great sense of gratitude for what I received and, little by little, I returned to the figure of Antigone, which I understood in a new way, as if my situation had something in common with hers. Except that what happened to the one was almost the opposite to the lot of the other. She had been deprived of the air, the sun, and all the environment necessary for living, whereas I was sent back to it, which had become worthless to my contemporaries. Their lack of interest in life itself is what saved my life—and my thinking.

I was also put back in the context of Greek culture, especially at the time of the passage from early to classical Greek culture, a passage that the tragedy *Antigone* by Sophocles precisely brings to the stage. Already, in *Speculum*, I commented on this tragedy in the chapter devoted to Hegel, in which I wondered about his manner of distributing the responsibility for human and divine laws between the man and the woman, in order to establish an ethical world that allows for a becoming of spirit. Besides, it was after proposing the figure of Antigone as a subject for a seminar that I was removed from my teaching at the University of Paris, Vincennes.

However, my understanding of the figure of Antigone evolved after the various events that followed the publication of *Speculum*. My interpretation was henceforth less institutional, be the institution the state or the family, and it started from a mere belonging to the natural world, as a place of dwelling or an environment, as well as an identity. All the cultural requirements, which aimed at overcoming, concealing, and forgetting nature proved to me to be more or less ingenuous devices for masking and hiding our real identity and sparing us its cultivation, each on our own part and also between us. Our traditional and current world appeared to me as a play, or the repetition of a play, in which life no longer circulated, nor grew, nor was cultivated or shared. Criticizing this theater, while asking nature to provide for my survival, was a possibility, to which I resorted during a certain time. But, this way, I was lacerated between two parts of myself, and, if some philosophers consider that such a tear is the very condition for thinking, it was not my opinion. I wanted to live and cultivate life, and thinking as a substitute for life did not suit me.

What did not suit me either was staying as a woman in a culture that was not appropriate for me, that is, agreeing with the fact that women's liberation could stop at a mere biological level. In fact, this would have amounted to perpetuating the status traditionally assigned to women, with the additional permission of going outside the family house so long as they submitted to a culture in the masculine that deprived them of the cultural values they needed for constituting a subjectivity of their own, or, in other words, for existing as women. Obviously, I could not accept such a choice, which, moreover, was not left to me. I was expelled from any establishment, including by women, and was not allowed to get a word in edgewise. Even Simone de Beauvoir refused to make an appointment with me. I had no other alternative but asking nature to teach me how I could preserve and cultivate life, beginning with my own life.

Thus, I had to return to a time even before the institutional conflict, in which Creon opposed Antigone, and inquire about the order the latter defended. If Antigone rebelled against Creon, it was not in the name of a subjective passion, not to say the caprice of a teenager who is not yet capable of understanding what governing the city required, as some valorous intellectuals claim. It is because she tried to maintain a natural order that the arbitrary laws Creon imposed on the citizens began to destroy. In reality, Antigone struggled for preserving life and the natural and cultural environment living beings need. Ultimately, it is not even a family matter, except for the concern about the familial order that Antigone attempted to preserve against another order, in which the family will be in the service of the state and those in power instead of remaining the place where life is cultivated.

The enigmatic and still impenetrable truth that Antigone represents henceforth dawned upon me as evidence for the expression of life itself. Antigone testifies that life is an absolute, the unique absolute that we have to preserve and to incarnate. No other absolute can supplant the absolute that life is without submitting us to mere survival. Passage to a culture of simple survival through subjection to values or ideals more or less arbitrarily constructed by men, Antigone opposes with a radical "No."

There is no possible negotiation regarding life: life is or is not. The question is how we can cultivate what life is. As life can never be treated as an object, our culture does not supply us with a method to answer this question. I had thus to clear a path that did not yet exist. And it was not by chance that, as soon as I opened a patch of sky or perceived an inkling of a new truth, all was made to cover such a light. I remember confiding, exhausted, in an old friend: "I am just emerging, I am buried again." I swear that I was not quoting some text or imitating some figure. I was only short of breath and trying to recover a way of breathing.

It is true that being born requires one to breathe by oneself. Instead of teaching me how to cultivate my breathing, my culture had taught me how to suspend my breath in words, ideas, or ideals—something that led me to breathe in an artificial way and left me breathless when I was expelled by and from my cultural background and indeed no longer believed in its values.

The death that was imposed on Antigone was to be locked in a cave, and so deprived of the air and the light of the sun that a living being needs. In order not to commit a murder forbidden by the law, Creon attempted to kill her without overtly killing, even ordering that food be brought to her in the cave. But how to consume food, how can the combustion of food happen, without air and light? Did Antigone have another solution to her imprisonment than taking her own breath away?

I was outside, and I could attempt to recover my breathing first to survive and, then, to discover how to cultivate life.

14–25 December 2013

3

SHARING UNIVERSAL BREATHING

At first, I had to recover life: a mere survival and also the life, of which my culture had deprived me. Unfortunately, I no longer lived in the country and no longer had a garden. Breathing is the first and the last gesture with regard to life, but to recover one's breath in Paris, the city in which I was living, was not an easy task. I left the city for the woods or the mountains as often as I could, in order not to stay in a house or a studio but to walk outside in open air. When I absolutely had to stay in Paris, I spent a part of the day in a park. The vegetal world was my favorite dwelling again, but, at least in the beginning, more as a place essential to my survival than as the Garden of Eden. Perhaps I could say that the vegetal world had become a mothering place that provided me with the air I needed. However, I was no longer a fetus. I no longer received oxygen with the maternal blood through the umbilical cord and the mediation of a placenta. I was already born and had to breathe by myself. It was the vegetal world that ensured mothering care with the environment it arranged around me. I could even re-form a sort of aerial placenta, in which I remained sitting for hours, and the trees or other plants purified my breath without asking for anything in return. I was very moved and grateful to nature for the hospitality it—sometimes I would like to say "she"—gave me always and everywhere without saying anything or calling for something in compensation. It-she only offered me a place where I could regain life and faith in life.

Together with the space and the air it provided me with, nature, above all as the vegetal world, gave autonomy back to me. It proved to me that I was dependent upon no one to live: breathing suffices for living, and I could do that only by myself, with the help of the vegetal world.

Furthermore, breathing is what allows for a passage from vegetative life to spiritual life. Thanks to the vegetal world, I could not only begin living again but also continue thinking. "There is air" was sufficient; I did not need another "there is." A new start and a new world were possible, without anything other than breathing, and so, little by little, opening in myself a clearing made of a reserve of free breath, in which I was capable of perceiving and shaping that which I perceived. At first, I could, above all, affirm "It is not that" and put in question the truth that my tradition taught me. I crossed through a sort of negative ontology. It is not that which has been passed on to me as truth that allows me to live and to think; the essential function of air and of breath has been forgotten as the mediation for both living and transcending a mere natural life.

This function of mediation, in fact, already existed between the vegetal world and myself. We were, in a way, in communion with one another. Air put us into living relations even if we did not assume the same role with respect to it. Through air, I participated in a universal exchange from which my tradition cut me off. Thus, I was alone and not alone. I took part in a universal sharing. Gradually, I experienced such an involvement, and this brought me comfort, gratitude, and also responsibility. I became a citizen of the world, first as an inhabitant of the earth who joined in a sharing of air.

In 1974 I discovered, or rediscovered, a crucial aspect of social and political life that my culture neglected: to take care of the atmosphere and of beings that maintain it. The cause was almost as difficult to defend as that of women's liberation. And, more often than not, I acted without saying anything: I stopped flying, driving a car, smoking, and I also became a vegetarian. All that did not happen from one day to the next, but step by step, and the most difficult step has been giving up smoking . . .

From the outside, such a universal communion through air might seem a fusional process. In reality, that is not the case. If breathing reminded me of the difference between the air in which I dwelt and the air that corresponded to my soul or my spirit, that is, of the difference between the outside and the inside, breathing also reminded me of the difference between the other and myself. Losing our identity to form a whole with the others, be they human or not human, amounts to giving up our own breathing, and this can lead to a terrible struggle for survival. Before any master-slave struggle at the economic and social levels, the struggle exists at the level of life itself for appropriating the air essential to life. However, we remain unaware of this or prefer to ignore the problem. And the presumed materialists barely take into account the necessities of our body, for which the first food is air. They misjudge the harmfulness of pollution in order to, supposedly, provide people with food. But who can eat without first breathing? Nevertheless, they do not hesitate to claim the reopening of the most polluting factories to get work for unemployed persons, factories, which, by the way, are the cause, according to them, of the alienation and the exploitation of the workers.

The contradictions of our systems, be they idealist or materialist, are, at least in part, founded on a lack of consideration for life itself, and its first and last gesture: breathing. It is meaningful that both Hegel and Marx consider so little the vital or spiritual necessity of air and breathing. Does not their thought address the dead more than the living? And does this not result in a humanity in which each asks the other for life instead of, oneself, caring about one's life at the physical and spiritual levels in order to be capable of sharing it with the other? What do our societies, unconcerned about air and breathing, look like?

If air is crucial for life, it is also essential as a fluid to ensure the cohesion of a physical and even a spiritual whole, be it individual or collective. If we were capable of forming every whole while taking air into account, our totalities would lose their systematic and authoritarian nature. They would also remain capable of transformation in order to enter into relations with an other, or to form a community with others, without each losing their singularity. A cultivation of our breathing, of our breath, allows us to be more malleable, to adapt ourselves to a situation without submitting to it. We, then, can endeavor to make our breath more vital or more subtle, staying alive but modifying our manner of being and acting according to the context. Neglecting the necessity and the potential of breathing, our tradition has rendered our subjectivity both weak and rigid because it is frightened of any change. As it has received its contours and forms from an outside world more than from its own life, it does not know how to deal with its presumed self.

In the woods, in the garden, I was contemplating the forms that a tree adopted, how it was able to change while remaining itself, a change in which it did not risk losing itself in devices because it amounts to the appearing of a living being. And I wondered why we, as humans, have ignored such an aptitude and thus resort to constructed forms to become acculturated. Why do we not keep alive and develop our own energy so that we may let our natural belonging flower? Remaining continuously with ourselves might allow us more plasticity to relate to the world, to the other, and be present to them more than when we are subjected to forms extraneous to ourselves, something that deprives us of autonomy and freedom with respect to any transformation.

Was not the incapacity to become aware of a new way of thinking at work in the rejection that *Speculum* provoked? Do we not consider, as Westerners, that war and destruction—or the so-called murder of the father?—are necessary to pass from one epoch of history to another?

Now, this did not amount to my purpose or desire: I wanted to come back to life and not to kill or destroy anyone or anything.

I thus decided to turn to another tradition, hoping that, this way, I could discover both help for surviving and a path to cultivating my life. It was a car accident that, first, led me to address a practitioner of yoga, as I tell it in *Between East and West* and *A New Culture of Energy*.[1] He was French, but he had done his training in India in the school of Krishnamacharya, and he was familiar with the Indian tradition. Beyond the exercises he gave me to do in order to develop my spinal muscles, he also taught me how to cultivate my breathing and recommended texts to be read and even masters to be met. Little by little, I discovered a universe that I was searching for in which breathing is crucial and life as such is respected and cultivated. I also found out that some masters—for example, Krishnamurti (cf. *The Years of Awakening*),[2] not to mention the Buddha himself—had asked trees for help, as I was doing.

I discovered a world with which I was already familiar, but without the cultural context that Eastern cultures henceforth provided me with. I began reading the literature of this tradition at therapeutic, cultural, and spiritual levels—for example, the Vedas, the Sutras, especially the *Yoga-Sutras* by Patañjali, the Upaniṣads, the Bhagavad Gītā, the Bhakti,[3] etc., as well as some novels and works of poetry, but also some good commentators such as Swami Sananda-Sarasvati, Shri Aurobindo, Mircea Eliade, Alain Daniélou, Lilian Silburn, Jean Varenne, Heinrich Zimmer, etc. More than entering into a new universe, I had the impression that I filled and populated my world with words, stories, images, characters, and even gods that suit a cultivation of breath and life. Perhaps I was rediscovering a sort of cosmos, as it existed in the early Greek culture, when *logos* and *cosmos* were not yet separated but formed a cultured whole? I could evoke the figure of Antigone again as opposing the division between nature and culture instead of pursuing the cultivation of nature itself.

Why ought this division to happen—and, somehow or other, has happened in diverse traditions? At what epoch of their evolution and in the name of what necessity did humans split off from nature, and from themselves as nature, instead of cultivating nature? What need forced them to accomplish such a gesture? Is it unavoidable?

I started inquiring about this evolution in my tradition, but also in others, and the presence or the absence of the vegetal world was often an indication of the time when humans began wandering through abstract elaborations and calculations aiming at mastering nature by means of technical logics or tools, instead of working toward its growing and blossoming, beginning with their own.

What has, then, become of the tree of Genesis and that of the Song of Songs, under which love would be awakened? Of the tree under which the Buddha was born, and also meditated, and Krishnamurti tried to find shelter and peace? Of the oak, from which the druids picked the mistletoe, or of the tree under which my friend dispensed justice in Africa? And of so many others . . . What has befallen these suppliers of air for our life and spirituality? Are the trees today nothing but material at the disposal of humans for their various businesses? What will happen to a humanity that behaves in this way toward its most precious common good?

28 December 2013–5 January 2014

4

THE GENERATIVE POTENTIAL
OF THE ELEMENTS

No doubt, *air* is the most essential element to terrestrial life, and it is also the element that has the greatest power to ensure a mediation between the different states of matter, within us and outside of us. It allows us to be both a body and a soul, and the atmosphere to undergo changes of density and temperature. As it is able to make possible the transition from the materiality of our body to the subtlety of our soul, air can provide for the transition from the earth to the sky. And if a lack of consideration for air in our way of thinking makes our thoughts rigid and immutable through a sort of drying, the lack of respect for the atmosphere transforms its fluidity into a glass roof.

Starting from a certain epoch, the continuity between the body and the soul, the earth and the sky, has been interrupted. Instead of ensuring the passage from the one to the other, especially through a transformation of the density of air, the two became opposed to one another, as if we ought to take care of each differently. Now it is not the case: caring about the atmosphere of the earth and worrying about the nature of the sky depend on the same gesture. In both cases the question is that of the gradual transformation of air itself. And the transition from our body to our soul obeys the same process.

When Mircea Eliade assimilates, as Tantrism also does in a way, the yogi to a plant, he reminds us of such a process that unfortunately human culture generally has forgotten, at least in the West. This culture, then, asks us to die to become spiritual and not to make our bodies flower spiritually. And the same happens with the sky. If we were aware of the fact that the sky corresponds to a sort of flowering of the terrestrial atmosphere, perhaps we would hesitate to cross and perturb, by flying, the various layers that lead from the one to the other.

The vegetal world contributes to maintaining the correct density and quality of the atmosphere, notably by acting on its level of humidity. And, for example, the consistency of the petals of a flower and its lifetime are dependent on both the vitality of the sap and the atmospheric conditions. In reality, the same applies to us as humans. However, we have pretended to master nature, as environment and identity, by valuing only what has a permanent and immutable consistency. Such a gesture has gradually substituted the appearance for the appearing of the living. Now, appearance has little to do with the quality of the atmosphere, whereas the latter is crucial for appearing. In a culture that favors appearance over appearing, the importance of air as a vital element is supplanted by the fixed forms of an image or an idea—one could say by "air" as only a manner of appearing. A sort of "looking like," which is approved and supported by social norms and fashions, henceforth substitutes for the flowering of a living being. Sociocultural surroundings, with various degrees of fabrication, are imposed on a humanity uprooted from its natural belonging. Hence, the vegetal world loses its value as the necessary environment for the growth of living beings and as an example of and for them, which is really a pity.

Beyond its essential use for life itself, air is also a medium that we need in order to receive and pass on all the sensory messages. It also creates a pleasant ambience, not too dense or too empty, so that we can stand up and move in it without difficulty. Providing us with an invisible dwelling wherever we are or go, air is also a faithful companion for the one who can pay attention to its invisible presence. I could also allude to the scents or the sounds it brings to us, and to the caress of wind as a gesture of love of the universe when we walk in the country. Despising our lack of air, how many things do we risk missing, including the word itself, that we pronounce or hear thanks to it! Perhaps, because of the stress we put on language, some of us, Westerners, continue cultivating air lightly through singing—saving, this way at least, air in the form of an aria or a tune. Obviously, this is not yet caring about the reserves of air that living requires. Rather, it is using them for one's own human advantage.

No doubt, some people will object to me that in the beginning is *water*. And various cosmogonies start with water. I made clear earlier that I was alluding to terrestrial existence and not to the existence of the living as an embryo or a seed. If it were the case, I would have begun with water or with earth. As humans, we live first within water, while the vegetal world begins its life within the earth. Whereas a plant will remain naturally rooted where it started living, it is not true for us as humans. Our first taking root is, in fact, relational: conceived by two, we start living in our mother and are linked to her through the umbilical cord. In reality, it is a little more complicated: the umbilical cord connected us with our placenta, which acted as a mediator between our mother and ourselves, notably at a hormonal level. The question of our own roots is, thus, complex, and this explains the numerous myths regarding our origin, but also our constant attempts to provide us with constructed roots in order to master that which escapes us as our natural commencement, given that we have to face both dependence and uprooting.

Our taking root corresponds to breathing by ourselves, something that we do not consider enough and is difficult to imagine. Have we not compensated for this difficulty with the hypothesis that we were born of the breath of God? Thus, born of the most subtle—as is air in comparison with other elements. How can we realize that we come into the world thanks to something so subtle? How can we assume such an amazing condition? In a way, born of the matter of the soul itself? I think that humanity has not yet understood its responsibility with respect to its destiny. It neglects this aspect of its specificity, which calls for a human concern, and not only a facet of one or another religion that finally reminds us of the importance of breath.

Perhaps, the first gesture of recognition, but also of gratitude, regarding our vital dependence on air would be to care about the vegetal world. We also ought to acknowledge the intelligence of our body that looks after our life by breathing, whereas our mind forgets, when it does not paralyze, the necessity of breathing.

Fortunately, our body also consists largely of water, and air arises from water.[1] This is probably the reason why most of the ancient myths situate water in the very beginning and define the original sin as the opposition to the flow of water that allows for the generation of the living. Then, the word for the liberation of the waters with their generative potential can be said as a "feminine goddess word," and it inaugurates the appearing of all, including that of the Father and all the gods (Ṛg Veda 10:125:7), whose role will be that of making, fabricating. There is not a unique God who creates the heavens and the earth: the elements that constitute the world preexist, and the question is one of supplanting the demons who prevent them from taking place in a living universe. Thus to create, above all, means liberating, letting be and organizing the world so that living beings can be, especially because water is not lacking.

Waters conceal in them the seed of all and are said to be "Mothers," that is, the place where the original golden egg, from which the world will arise, is generated. Thanks to the ardor of their desire, the original waters are even capable of generating light and *fire*—and so giving birth to the god Agni (Ṛg Veda 10:121:7). The sun, the day, the dawn, and fire appear as soon as waters flow out, bringing the golden egg or embryo that diffuses light as the sun, fire, or gold. And the gods, at first Indra and

Viṣṇu, will have to construct a clearing for the light to shine on the earth after the egg had been broken.

Such an account of the origin of the world looks rather similar to our own begetting as humans, except that the semen intervenes only in certain versions of the myth. The most surprising aspect for us, as Westerners, is that the power of generating is entrusted to the elements themselves and not to the men or their God-the-Father. It is the evolution of the elements, and, in particular, the combination of ardor with waters, that gives birth to the whole universe. The role of the masculine gods is to technically build what has been generated: to consolidate the earth, to extend space, to shore up the sky, to make the rivers flow. One of their main actions is to lay out the sacrificial area which is the place where the perpetuation of the cosmic order is cared for.

Surprisingly, the word for designating this fabricated space devoted to worship amounts to the English *clearing* as a free space opened in the woods, in order to let the light of the Heavens shine without the hindrance of treetops. How could one avoid associating this Vedic clearing with the Heideggerian clearing?[2] And also noticing the absence of allusion to air in both contexts and the role of trees regarding this element? The wood of a felled tree ought to serve as the pillar that will ensure the separation of the heaven and the earth, and so arrange an intermediary space between them. The forgetting of air, as providing a gradual transition from the earth to the sky and occupying the intermediary space between them, conveying light from the one to the other, is astounding! Only later, it will be spoken of the wind, but not of air as such. And if it is specified that living beings must eat to survive and develop, there is no mention of the necessity of breathing.

Now, besides the fact that breathing is our first gesture of coming into the world, it is also a gesture that can define our internal and external space or place. Breathing is an organizer for living itself and for the coexistence between living beings. The cosmogonies do not consider the crucial principle that air is a basic element for the cultivation of life itself. Must we read that as an indication that humanity as such is not yet born? Not yet living on the *earth*?

The masculine gods resort a lot to their ardor to put a world, their world, in order, from the elements, but they barely take into consideration the potential of these elements, especially their structuring potential. It is true that two versions of ancient cosmogonies exist in the Ṛg Veda. In the first, which corresponds to a peaceful genesis, the waters—sometimes said to be "Mothers"—thanks to their ardent desire, generate alone the original egg from which the world will arise (10:129:3); in the second, where the hero is the warrior god Indra, the creation of the world results from the warlike deeds of masculine gods against the guardians of the original waters. And one of the most important events in the passage from the first version to the second is what could be called the betrayal of Agni, the god of fire, who swears allegiance to the clan of Indra, not without shrinking back and feeling ashamed, with the aim of receiving immortality. Thus, Agni, who is first generated by the ardor of the original waters, gives up having a share in their world to be in the service of Indra, especially for all that concerns the sacrifices. Instead of remaining attached to the potential of the elements, Agni enters the service of Indra to contribute, as fire, to the sacrificial rites. It is no longer the elements themselves that provide ardor, sap, growth, and the appearing of all, but the work of fabrication of men, initially of masculine warlike gods, which gradually attempts to substitute for the potential of the elements. At the same time, the "Waters" become "rivers" and, as such, the submissive spouses of Indra, who sing praises of him.

Before any theft of fire from the gods—see, for example, the story of Prometheus—the fire is, first, stolen from the original Waters, also called "Mothers," that is, from the generative potential of the elements. But, cut off from its natural source, fire loses its regulation, as the whole is deprived of its own growth. The world becomes subjected to the fabrication of man, who "sometimes accomplishes evil, sometimes achieves brave deeds" (Sophocles, *Antigone*, 65–66). Man, anyway, transforms a natural energy into a more or less artificial energy, which both lacks its own resources and subjects the whole to fabricated ends and forms. The becoming of those who or which grew by themselves with their own forms is interrupted. And, perhaps, only the vegetal world— and some wild animals?—henceforth continue bearing witness to the potential of an elemental origin: rooted in the earth, growing in the air thanks to the light and warmth of the sun, but also the humidity that the rains bring. Following Aristotle (in *Physics* B1) one could say that finally only vegetal life testifies to the potential of *phusis* itself.

The status of fire then becomes ambiguous and problematic. Man's exploitation of man begins with that of the energy of all living beings, arranged in an intermediary, fabricated world, between the two parts of the original broken egg: one serving as the sky and the other as the earth, two parts between which elemental energy no longer circulates on an earth that, in fact, is still missing. Might one say that the trees from then on remain the privileged surviving witnesses of the past energy of the living? Hence, their importance to the ones who try to revive the link between a mere survival and a new becoming of the world, beginning with that of the human being itself.

7–18 January 2014

5

LIVING AT THE RHYTHM OF
THE SEASONS

The elements do not easily let themselves to be mastered by man. They resist human domination as long as they can, notably in the form of the seasons—which, unfortunately, are disappearing today . . . Indeed, the seasons bear witness to a passage from one element to another, to the interaction between them and their transformation into one another, which Empedocles alludes to still. Preventing this continuous transformation from happening paralyzes the whole and renders it infertile.

At the time of the Vedas, the daemons—sometimes compared to dragons—keep the waters confined and prevent them from flowing out, so ensuring the generation of the world. One could say that the current imbalance of the seasons, which results from a diabolical appropriation of the elements by men and their reduction to objects at men's disposal, is also rendering the world infertile. We commonly hear people saying with sadness, "There are no seasons anymore!," whereas to allude to seasons in cultural circles seems out of place and looks outmoded and artless. A radio journalist even criticized me for talking about the seasons, as if it were something politically suspect! I imagine that the same journalist broadcasts on the occasion of Christmas or Easter, for example, but without establishing a connection between these feasts and cosmic events that they, in reality, celebrate.

There have existed endless debates, some of which Eliade alludes to, about the precedence of natural productions over divine creation. And many people, especially women, have been sentenced to death because they upheld the precedence of nature. It is really surprising that it has not been possible, at least in our tradition, to harmonize the power of nature with the divine power. This has divided us into a natural part and a spiritual part, instead of working on a solution of continuity between the two parts, which is absolutely crucial to accomplishing our human life. Moreover, the natural part has been assigned to women and the spiritual part to men, so much so that they can never really marry one another. Now the sacred marriages, also called hierogamies, were decisive events in the generation of the world. They must happen between two elements or divine figures that are considered to be of equivalent value, and if the woman, or the feminine, is undervalued in comparison with the man, or the masculine, then no union between them can happen, be it at the elemental, divine, or human level.

All these stages are initially based on a cosmic dimension that gives meaning to them, a thing that our contemporaries understand with difficulty, reducing the whole to a human aspect. We are a long way off from the epoch where love between gods was presumed to act on the weather! Now, we find mentions of that in many texts of various traditions—for example, about Śiva's love stories, which can either make the sun shine or bring lightning according to whether the lovers are harmoniously embracing or come into conflict.[1]

We can also observe that the amorous unions, as hierogamies, do not happen at any time of the year but in the *spring*, and that they are immersed in a vegetal environment and even compared to what happens to the plants. In the Song of Songs—which takes up many themes, and even words, from poems regarding the more ancient Sumerian hierogamies[2]—the lover invites his beloved to awaken because the winter has passed, and the garden is the place where they will meet and embrace each other. In these poems or songs, we notice a sort of scenography that presents the relationships between agrarian, pastoral, and regal warlike epochs, which are more or less favorable to the flowering of the weddings. Indeed, they allow, or disallow, family ties—especially those with the mother on the side of the bride and those with the father on the side of the groom—at the same time as they gradually move away from a cosmic hierogamy, which is their very origin.

There is no doubt that the closer we remain to cosmic hierogamies, the more the weddings are both carnal and divine. The tearing or the rift between these dimensions has not yet happened, and the lovers enjoy sharing their natural and spiritual belonging, which brings them energy, happiness, and fecundity. We can also note that, at that time, she takes part in the caress as well as he does, and that the male body is present and celebrated as the female body, without focusing on his sex, in a way extrapolated from any embodiment.

The renewal of the year happens in the *winter*, when the light returns and the days are getting longer. That time corresponds to the celebration of birth—the birth of the light and the hope it gives us that life will spring up and grow again and that love will become possible anew. A god can come into the world at the winter solstice, but it is too early and too cold to marry. Love can only be foretold—why not in the form of buds?—but not yet fulfilled. It needs more warmth and a light that is of a solar yellow and not only white. The sacred marriages happen in the *spring*, when the heavens and the earth unite with one another through a still fresh air and sunshine.

Then the atmosphere can be really divine and the earth itself—let me say: herself—seems to smile through the new leaves and the slow opening of the first flowers. The birds sing. And their song resonates in a deep silence that appears as a wonder beyond any word. The air looks clear, but it is nevertheless crossed by countless invisible vibrations, bonds, presences, that can be perceived and create a cosmic energetic communion, the density or nature of which is light, virginal. The colors in the garden are still bright, above all white and pink. The whole grows, but life, as love, remains discreet. It is a time of betrothals, shared by the entire world, in which all are in search of the way of celebrating love: moving forward, flowering, singing, strutting about, dancing, but also dreaming and imagining a new way of approaching, gesturing, speaking, or keeping a meditative silence tuned into the amorous mystery of the living. The spring is the time of a universal falling in love, of experiencing the rebirth that then happens for each with the, often clumsy, impetus for sharing it, passively or actively. Undoubtedly, the spring is the most wonderful and divine season that we have failed to cultivate and that is more and more vanishing.

The West has favored the *summer* to the detriment of the spring. This probably results from a religious tradition that does not very much consider the spiritual potential of carnal love. We do not find in the West, as is the case in India, lovers hugging each other on the frontispieces of temples—for example, of Konarak or Khajurāho—or a sacred book such as the Kama Sutra. We neither know the treatises on the pillow that the lovers consulted before making love, nor are we familiar anymore with the beauty of Japanese etchings, depicting spring, contemplated by the woman while her lover embraced her. Sexuality in our tradition always seems to be an infringement, and not the place of a possible spiritual becoming and sharing. However, without cultivating our sexual energy, what becomes of love, of our body and our soul? Are—or were—not the traditions, in which humans can imitate their gods by innocently making love, in harmony with the rhythms of nature, more sanctified? As Aeschylus writes, "The holy Heavens are wild when they penetrate the body of the earth."[3] Are not these words more enthusiastic and supportive than our medical treatise—as useful as it may be—on *The Woman's Sex* by Doctor Swang, or our more or less saucy and licentious erotic literature? Why did we come to such a stage, in which the philosophers who venture to speak about sexual relations—including Sartre, Merleau-Ponty, and even Levinas—use words that express more a master-slave struggle than a divine amorous sharing?

The West has focused on the fruits more than on the flowers. And carnal love has become the way to produce children and not make lovers flower. In reality, the *summer* and even the *autumn* are preferred to the spring in the West, probably because our culture is a masculine one and the stress is put on production, through which man can compete with nature. Instead of celebrating and helping the natural growing and blooming, he endeavors to do better than nature itself through his fabrication (e.g., Aristotle, *Physics*, B1). He tries to master the natural potential to substitute his own power for that of nature. He thus paralyzes the natural life and production by adapting them to his own plans instead of letting nature grow in accordance with its rhythms and its fecundity. More and more, he strives to intervene in this process, to force nature to produce more and at an accelerated artificial rhythm.

No doubt, the time of fruits—conceived in a general sense as the fruits of the earth—is also a beautiful time, and the earth, in its great wisdom, indicates to us what it is advisable to eat at each time of the year. For example, it is not by chance that potatoes, walnuts, and chestnuts are harvested in the autumn: they respond to what our body then needs. Unfortunately, there are our caprices, or those of our society, which decide on our way of eating without us paying attention enough to what nature provides us with, and giving thanks for it. It seems that man is henceforth incapable of glorifying nature for the life it lavishes on him, and he wants to prove the superiority of the strength of his work in comparison with natural fecundity. He will produce by himself, and not contribute to the development of the productions of nature itself. All that amounts to a loss, instead of an increase and cultivation of energy, that harms all living beings.

After the pickings and the harvests are finished, our tradition celebrates the day of the witches, Halloween, and the day of the dead: it is not a matter of chance that the two celebrations happen with an interval of only two days. The end of October and the beginning of November mark the onset of the time of nightmare, of magic and irrationality, and also of death. We are entering into the season—*winter*—in which the nights are the longest and almost nothing appears on the earth. Now, our culture has favored brightness and visible productions. It does not value, but even fears, the dark, or the hidden, and it refers the invisible to the absolute light of God—as an exorcism? The secret germination of plants and even that of a human being are not assessed as it would be worthy of them. They bear witness to the fecundity of the sap in the darkness of the soil, or of a womb, and to the fact that appearing amounts only to a part of the growth of life. Scorning this secret process of the living risks mistaking appearance(s) for the appearing of life, a risk that lies in our tradition from the very beginning and that has transformed the West into a culture of uprooting.

For a long time, women remained closer to a world of roots. And, especially, those who have been called "the witches," who remain faithful to the natural world, the living and curative properties of which they know. They do not fear the winter because they share terrestrial energy and can feed themselves on the roots that the earth produces at that time. Alas! we heard or read what happened, and still happens somehow or other, to these women faithful to the earth (e.g., the beautiful novel *Corrag* by Susan Fletcher)."[4]

Nature offers us a wonderful place in which to dwell. One and unique, it is also always changing and becoming, according to the seasons and the geographical place. It provides us with all we need at each time—to breathe, to eat, to contemplate through all our senses, and also to share. How rich is the rhythm of the seasons in comparison with our calendar!

Our annual schedule looks so abstract, gray, ruled by money, in relation to the uncountable variety that the seasons bring to us. Nevertheless, we continue alternating work and rest, the time of shopping for the feasts and the time of sales, and so forth, rushing from one to the other without finding a moment to contemplate our natural environment, without making time to enjoy, cultivate, and share life. And so life, little by little, vanishes—ours, that of our planet, and that of all living beings that inhabit it.

23 January–6 February 2014

6

A RECOVERY OF THE AMAZING
DIVERSITY OF NATURAL PRESENCE

Participating in a universal communion through air, bathing in the generative power of the elements, living at the rhythm of the seasons, transformed me, little by little. I experienced what it means to be alive as well as to meet with other living beings. I surfaced from a world in which all is planned in a more or less arbitrary and imperative way. I discovered a real freedom that can be gained only by oneself and with respect for others as living. I was no longer, at least partly, masterminded by an order that had nothing to do with me, nor did I merely oppose it. I tried to respect the rules essential to coexisting in society, but as a structure that did not subdue my own subjectivity. They remained external to it, as a sort of more or less suitable civil context that I had to take into account in order to move in the city without losing energy in useless conflicts or discussions. The important thing was not there. Anyway, all that was already dead, and my desire was to live and elaborate a culture of life.

I was thus there and not there, in a way absent, as they constrained me to be. Unless I met a living being: a little child or, more seldom, an adult; a bird or some wild animal; a tree or a flower. Then I entered into communion with them. Those were festive moments, during which I shared much more than superficial and conventional words. Life happened and flowered, if only for a moment. Sometimes, it was the gift of the day, which contributed to my persevering in making the way. I began writing a poem each evening, as an everyday prayer,[1] to celebrate such events, which, otherwise, would have sunk into oblivion. Now, they were crucial runway lights to open up the path and help me to pursue it.

In reality, being alive means being rooted in oneself, but also being vulnerable. Most people perceive very little, and become almost insensitive, except to what advertising recommends them to feel. They survive with a sort of armor, in which only a little window is left open, which allows only for selective perceptions. As I came back to life again, I was feeling many things through my entire being, which corresponded to an incredible richness, but also to a danger. How to experience all I was experiencing without losing my own subjectivity and without suffering because of the brevity of the meeting, either? I thought of that after encountering a little child whom I did not want to leave unhappy with our separation—but the same was true for myself . . . How to succeed in unifying so many experiences without depreciating one in relation to the other or vanishing into such a multiplicity? How to organize them without bending everything and everyone to a whole, to my whole, as man did in Western culture? I did not want to favor our traditional intelligibility over sensibility or sensuousness, or the relationships to an object, be it an Idea or an Absolute, over relationships to, and above all with, other living beings.

I had to turn my culture upside down, or to reverse what my culture taught me. However, when it is a question of life, this cannot happen from one day to the next, and, instead, required me to advance step-by-step. What allowed me to act in this way was staying each day for a time in nature: in a garden or in a park, in the woods, in the mountains, or by the sea, depending on my availability. This helped me to regain both my strength and my unity, but also to clear my path. Besides, it was an enchanting moment, which supported me for the rest of the day, and beyond. In many of my everyday poems, I evoke something of the enchantment that nature brought me, an enchantment that humans too often ignore, whereas the birds sing it. As they do, I, myself, also tried to express my gratitude through words, as close as possible to the natural microevents I attempted to celebrate.

What nature offers us to experience is so beautiful, multiple, and speaking to all our senses that, more often than not, I prefer to spend a moment in it, instead of going to the cinema or visiting an exhibition. If I continue doing this, it is above all to take the works of artists into consideration. But, rarely, do these works provide me with as much life, energy, and joy as nature does. Instead, they often take from me the energy that I accept to devote to them in recognition of the works of the creator.[2]

When I stay in nature, I am in a living environment. I thus receive and share living energy. What is more, nature is a kindly magician. For example, in the autumn, when the sun disappears, the leaves of some trees become yellow and this way create a solar atmosphere in the garden. In winter, the branches of the trees and the bushes are already covered in little buds that pass strength on to me and the hope of springtime—another proof that the winter is not merely a time of death for the earth. The forms and the colors of every season look suitable for making harmonious surroundings. The leaves are small and of a soft green color in the spring so that the light and warmth of the sun can get through, whereas they are thicker and dark-green in the summer in order to protect us from a too intense light and heat. The fruits of the spring and the beginning of the summer are very tasty and of a fresh red color, whereas those of the late summer or of the autumn are more nourishing and of a darker color. The first have almost no skin and can be eaten as soon as we pick them; in the summer, the fruits become covered with a skin that it would be better to peel to savor them; and, in the autumn, some of them finally have a shell that we must crack to eat them.

All this happens discreetly. It is wonderful and can seem a miracle, but most of the people consider all that to be normal, owed to them, and to be consumed without praising nature for its fertility or giving thanks to it for all it-she lavishes on us. In the past, it was common to recite a prayer before and after each meal, and such a gesture was, perhaps, the best way to establish a community. Alas! the celebration of the fruits of the earth was too quickly replaced with an invocation to God and then given up, apart from certain places of worship, where allusion to them exists, but really briefly and without being the true motive for gathering together. It is a great pity that things have turned out this way, because the celebration of sharing together the fruits of the earth is a suitable manner of defining a common space and time.

Obviously, the festive meals organized around the consumption of an animal cannot be endowed with the same meaning. They do not amount to a sharing of the space and the time that are determined by the natural rhythms, with their respective fecundity, and, furthermore, they presuppose the killing of a living being. If the meal can correspond to a community celebration, this is no longer a potentially universal community, but, instead, the meal marks the particularity of one worship or one clan, even with possible warlike connotations. In reality, it is closer to a sacrificial ritual or commemoration than to a peaceful and joyful giving thanks for the fruits that the earth lavishes on all of us.

I alluded only to some examples of the diversity of the gifts our planet provides us with. I spoke too little of the infinite variety of the views or the spectacles nature offers to us, and of their perpetual evolution, so that walking or staying in nature is never dull if we pay attention to our environment and accept being in communion with it. Contrary to what can happen with a constructed universe, the diversity of a natural world does not lacerate but, rather, unifies us. This seems to be another mystery that could be clarified by the fact that each, in such a world, is rooted and faithful to its roots, letting us live and even sending us back to our own life, instead of using it in order to establish its own existence. Each remains itself, and each proposes a sharing without infringing on the life of the other. And, in such surroundings, every perception communicates living energy to us, if we agree to receive it.

What also matters is that in the relationships with a natural, and, especially a vegetal environment, quality prevails over quantity. In our Western logic the latter has been somehow or other privileged, and this did not favor a cultivation of life. Our culture seems to behave in relation to the living as the first warlike god of the Vedas in relation to the elements. Too often, Western man has tried to grasp natural sap or energy in order to subject it to his own constructions through quantitative evaluations. Now, life as such has more to do with qualities than with quantities, above all when these are defined by forms foreign to it. Life develops with its own forms, and substituting more or less arbitrarily defined forms for them tears it apart and both lets it flow toward nothing and paralyzes its growth. Unfortunately, our cultural tradition acted in this way, and its appropriation of sap and vigor did not end in the growing and blooming of living beings, but in an artificially lifeless culture in which they struggle, especially against one another, for their survival through a blind search for quantities, notably of energy, that they cannot use and that they expend rashly and thoughtlessly. This amounts to a really irrational way of acting because nature brings plenty of energy to each of us, so long as we give up appropriating it in order to subject it to our own aims. The tree that I meet when walking in the woods passes much energy on to me, whereas it is not the same with a table that is manufactured with the wood of a dead tree. No doubt, this material is better at an energetic level than a completely fabricated one, but it is already cut off from its sap, and its forms are planned by us.

Agreeing to receive and share energy with the elemental and vegetal world is an inexhaustible resource of energy with living modulations and regulations—as can happen in loving relations between humans who are naturally different.

7–14 February 2014

7

CULTIVATING OUR SENSORY PERCEPTIONS

Our tradition has used our sensory perceptions as a means of appropriation toward a culture extraneous to a cultivation of life as such. It did not consider our sensory perceptions to be possible ways of entering into communication with nature, with the other(s), and with ourselves. As a mode of exchanging, our tradition resorted to words, which are presumed to represent the real, and to their gathering into a *logos,* leaping over the potential of our senses. Thus, instead of lingering before a tree—or a flower—to contemplate its singularity and meet it in its reality, we pass it, at best, thinking: it is an oak—or a daisy. We integrate them into our own world, sometimes wondering about what we could do with the fruit or the wood of a tree, or picking a flower, removing it from its roots and its own growth. We fail to meet another living being and to welcome the benefit this could bring to us.

Our culture taught us meeting a tree only through a denomination, an idea, a use, or a "face" of the tree that does not move, so renouncing both a great part of our present *sight* and the energy that an encounter between living beings can procure.

In reality, a tree—as any living being—is not only seen by us, as is the case with most of the fabricated objects; it also gives us to see because it lives by itself. I could say that looking at a tree brings me energy, whereas looking at a manufactured object takes energy away from me, notably because the first unceasingly creates the space and the forms of its appearing, whereas the latter needs me to re-create its possible place and its appearances.[1]

A tree gives us back our potential of vision, brings us back to ourselves with a capacity of seeing and of living, of which we are deprived by most of the familiar objects that surround us. Indeed, these impose on us a certain way of looking at them especially according to their use, as Jean-Paul Sartre analyzes with regard to the "tin opener" or the "corridor of the metro" and, more generally, our technical surroundings.[2] Such surroundings, which are now usually ours, can give the impression of our being the masters of a universe that we have created and dominate. However, more and more, they hold us in their power and deprive us of our living energy. They gradually weaken the potential of our senses, reducing them to a mere means of recognizing what we meet as something made, or at least coded, by man, and not as an autonomous living being.

Sensory perceptions then become dependent on human ideas or plans that cut them off from their living roots and qualities and remove them from their capacity to build bridges between our bodily and cultural belonging. It is through paying attention, in the present, to its concrete singularity and its sensible qualities, without substituting a name for them, that the perception of a thing, above all of a living being, can lead us from a merely physical stage to a spiritual stage of our concern. Gazing at a rose can help me to achieve a concentration that many words or discourses are not able to grant me. The combination of the sensible qualities of the flower gathers me, thanks to an attention they awaken at various levels, and, imperceptibly, I am brought from concentration to contemplation. If I take the time to live such a state, it can be converted into a sort of ecstasy, which results from a culmination of energy.

No education introduced me to such an experience in Western tradition. It is from Patañjali, and his *Yoga Sutras*, that I came to know the energetic and psychic profit of cultivation of my sensory perceptions.[3] However, for Patañjali, the way to develop perception toward a culminating energy—so as to reach *samadhi*—requires the mental internalization of what is perceived until the abolition of the subject-object duality. As for me, I think that the state of *samadhi* can be attained by preserving the duality between another living being—an oak or a rose, for example—and myself, but changing my way of perceiving. I no longer gaze at the tree or the flower as a solely visible object, but I also gaze at their invisibility: the sap which animates them and starting from which they can appear to me without any possible appropriation on my part, be it physical or mental. I think that it is the irreducible duality between the tree or the flower and myself that can lead me to a state of *samadhi,* arising from the love of life itself.

Obviously, such a path to *samadhi* is still more crucial when what I perceive is another human, especially a human different from myself. Furthermore, I can then achieve a sharing of the state of *samadhi,* which means a more accomplished fulfillment of desire. And this can happen in carnal love as well and make of our sexual attractions a path toward a spiritual becoming—toward a state of *samadhi* and enlightenment in two, more intense than it could be for only one.

But I am still alone in nature and I will, perhaps, know such an experience later, after returning among humans. I only glimpse its possibility from what I already know through a cultivation of my energy. However, the latter requires me to change my way of perceiving other living beings, beginning with the vegetal world. And this, little by little, will modify my manner of seeing the world, will transform me into a human who coexists with all the elements of the environment without aiming at dominating them. I, then, live with, share with, exchange with them, each of us bringing to the other and to the whole what the singularity of our being allows us to bring, but also receiving from the others what they can provide us with.

Then man is no longer the one who gathers everything and everyone in a whole, starting from himself and his language: a human is a living being among other living beings, each one remaining faithful to one's roots and natural belonging without mastery over or confusion with others. It is the specificity of each embodiment and its respect by others that secures and maintains the place of every living being.

I first talked of the necessary education, I could say "conversion," of our manner of seeing, because sight is the most decisive sense in our tradition. We must learn how to look at a tree, not to perceive its present form in order to re-present it mentally and fix it by naming it. Rather, we must gaze at its being as living and changing. Now we designate a birch with the same name in the spring, the summer, the autumn and the winter, although this name refers to forms, colors, and even to sounds and to odors, which are absolutely different according to the time of the year, not to say that of the day. Using the same name to allude to the birch at any time, we remove it from its living presence and deprive ourselves of our sensory perceptions to enter into presence with it. However, is it not the mode of presence that our culture taught us to consider the truth?—a truth that asks us to give up our living perceptions. How could we, then, care about life, ours and that of the world? Must being cultivated mean becoming unconcerned with life itself or, instead, learning how to cultivate it?

As for *listening*, what can listening signify if it does not imply a relation to, and with, a living being here and now present to and with me? To reach such a way of listening, we must first experience what silence is. Silence is the origin and the medium that allows us to listen to another living being. Our tradition has favored speaking to the detriment of keeping silent. I could here contrast Hegel with Buddha. For the first, the end of the path is being capable of gathering into a whole all the possible discourses; for the latter, it is to become capable of reaching silence. In our tradition, silence has been left, with undervaluation and even with contempt, to nature and to the women assimilated to nature. It is true that, without the capacity for keeping silence, I cannot meet with a tree or a flower. And it is also true that a woman must be silent, or use a phatic language—as Roman Jakobson said—to provide the fetus, or the newborn, with a free space for its existence and its growth. She is more accustomed to listen to another living being, and not only to a coded message, a gesture that is also necessary when it is a question of vegetal being.

Listening to the uniqueness of another existence, and considering its irreducibility with respect to my own, is a way to overcome the dependence on a truth, a discourse, or a master presumed to know the whole. It is to recognize another life as transcendent to my own and to my world, forever unknowable to myself. Thus I listen to this life, letting it be and grow, as to something that I cannot fabricate or master.

Preserving a space-time of silence between another living being and myself also maintains the possibility of a return to myself, within myself, by protecting our respective limits. No doubt, my breath is not totally independent of the life of plants; nevertheless, I must be faithful to my being a human in the relations to and with them: for example, sometimes bringing water to them so that they can provide me with pure air.

Silence is crucial for a being-with, without domination or subjection. It is the first dwelling for coexisting in difference. It is, or it creates, a place where we finally can listen to the other, and not only register a message as a machine could do. Then, I can listen to the music of the wind in the leaves, and also to the sound of the wood in accordance with the warmth, the dryness, or the level of humidity of the atmosphere. And all that contributes to restoring to me my breath, my freedom, and a living presence, in spite of so many codes that had transformed me into a robot.

I lingered again on hearing because it is the second sense that our culture appropriated toward constructing its theory through *logos*. I have not yet approached *taste* and *smell*, which our tradition too quickly left to mere natural belonging, to the level of our needs, without considering their human and spiritual potential. Whereas in certain Eastern traditions a god corresponds to each of our senses, in ours, ascending toward God often implies renouncing the sensory power of our perceptions. Enjoying the *taste* of food is rarely assimilated to a religious gesture in the West, but in ancient Eastern worship it can mean celebrating Viṣṇu, the god who creates and restores life. I experienced how much this intuition is relevant, and I recommend to my contemporaries that they learn how to celebrate Viṣṇu instead of taking medicines. Obviously, the matter is not one of swallowing a lot of food or of alcoholic drink. Rather, it is one of choosing really good food and drink and enjoying them, while praising nature, the humans, and the god who permitted us to experience such wonderful taste. This gesture can make of taste itself a path toward a spiritual becoming.

Smell is, above all, considered by us to be an animal sense, and to read the examples that the dictionary *Le Robert & Collins* sets at this proposal is revealing. When it is not a question of the smell of a dog, almost all the references to the use of the word are to bad smells, unless smelling takes a figurative meaning. The authors seem to have forgotten the ancient importance of smell, including that attributed to the god(s), in the religious rites. Only some yogis, and the man of the mountains that Nietzsche was, still remember smelling as a path for spiritual becoming. Now our nose is certainly the privileged sense to pass, above all through breathing, from a natural life to a spiritual life.

And *touch*? Educating our sensory perceptions, as I suggest we ought to do, involves reawakening the sense of touch in all of them, notably by accepting not to grasp but to be touched by the sight of a birch, the song of the wind in the woods, the scent of a rose, the taste of a raspberry or of a peach. There are also the warmth of the sun in which we can bathe, the breeze that caresses our skin, the various kinds of contacts nature offers to us . . .

However, it is perhaps touch that encouraged me to take the risk of returning among humans.

2–9 March 2014

8

FEELING NOSTALGIA FOR
A HUMAN COMPANION

The desire to have a human companion again has happened in the garden, or in the woods. As I wrote in a poem on 29 March 1998, "Only in the garden / Could you arrive. / There, the fullness already exists / And happiness. / Your presence will be something more, / Tasted for itself, / Non-confused with the whole. / The human appears /"[1] Feeling desire for a human companion cannot substitute for the need of maternal or medical care. It must arise after acquiring autonomy and recovering health. Sexual attraction cannot merge with a mere sexual instinct or drive, either. It implies that we long for something or someone that transcends the fullness we can experience alone in nature. It involves access to a new state, to another stage of the achievement of our humanity that we cannot reach outside of our relation to and with a human different from ourselves.

Obviously, longing for the presence of a companion can occur at any moment, whether this companion already exists or is only wished for. However, I remember two occasions when my desire was both really pure and intense. And each time this happened when I felt the fullness of a communion with nature. The first time, I was in a wonderful garden on the twenty-ninth of March, that is, in the beginning of spring and, the second, I was in the woods on a day in August. It was the perfection of what I lived that awakened the desire for a more complete sharing—as a celebration as well.

Ought our life in the Garden of Eden to have been this—a celebration of nature, including our own, as a divine gift, before pretending to appropriate the knowledge of good and evil? Was the matter not one of sharing happiness instead of longing for passing judgment on all?

Passing a judgment is a way of transcending mere natural life by cutting oneself off from it, without making life blossom as human. Could this be the path followed by Western man because of an inability to share in difference, which requires another access to transcendence that is provided by nature itself? In the former perspective, becoming human asks us to renounce our natural energy in order to enter a constructed world; in the latter, it needs cultivating this energy toward its humanization and its sharing with respect to any other living being, beginning with a differently sexuated human.

Experiencing fullness in nature, I was unwilling to give up such an experience for a disembodied transcendence. Indeed, this sounded to me quite similar to the original sin: the first couple wanted to appropriate the divine power of deciding on good or evil instead of enjoying being together in a beautiful garden that gave them all they needed and could desire. In reality, I think that more often than not we continue taking this wrong path. I intended to try starting from a natural fullness to build a future on the acknowledgment of this fullness. How could I succeed in that?

I began to write a poem each day in order not to lose the memory of the happiness I experienced in nature.[2] Such a practice already modified my everyday way of feeling myself and having my life thanks to a thread and a flow of bliss. Imperceptibly, then, happiness in nature came to mingle with what I felt, or what I wanted to live in love. I perceived with a fresh comprehension that happiness in nature and happiness in love could, and even ought to, interlace with one another and that I had to clear a path that allowed me to combine the two. The words *sensible transcendence* occurred to me as saying both a longing and an intuition, that is, a global, embodied perception likely to open a new horizon for my human journey.

As I have been educated in the context of a Christian tradition, the experience of a sensible transcendence ought to have been familiar to me. But the manner of transmitting the good news of the Gospel did not allow me to know what the sensible transcendence that Jesus could represent meant. Unless I attribute too much to the Christian message: What does the divine becoming man mean, if not the possibility that transcendence exists in our humanity? Perhaps the question is whether the passing on of the Christian legacy has remained under the influence of Platonism or whether it has been mistaken in the same way as Adam and Eve were in pretending to reach divinity before accomplishing humanity, that is, to leap over the cultivation of our incarnation. And Nietzsche was not so mistaken when he said that only one Christian has existed, or when he proclaimed, though the cries of a supposedly mad man, in the public place or in the church: "'Where has God gone?' . . . 'I shall tell you. We have killed him—you and I. We are his murderers. But how have we done this?'"[3]

Anyway, the stress put on paternal genealogy in the transmission of the divine belonging of Jesus prevents what I call sensible transcendence from happening. I think that such an experience can occur first on a horizontal level and between two differently sexuated bodies. It needs the presence of physical matter and the sexuate belonging that both animates the body and puts it in relation with another body, while entailing transcendence when the two are not of the same sex and are faithful to their natural belonging.[4] Sexuation is not only an empirical and secondary thing with regard to our being; it is what brings to it a specific morphology and individuation. Before any sexual practice, our identity is sexuate, unless it is a mere abstract construction that cannot be shared with the vegetal world. Remaining in a living communion with plants, while assuming my human condition, required me to embody the transcendence of my sexuate belonging. Longing for a human companion signified such a desire as well as such a necessity.

Nevertheless, tearing myself away from the bliss I lived in nature was not so simple! But did I not have to pursue my human journey, also in order to respect the plant world, which welcomed me, mothered me, gave me so much felicity? Had I not to take it as a model of perseverance in cultivating life: in preserving it, in making or letting it grow and blossom, in sharing it? Could I be satisfied with enjoying the happiness nature brought me? Anyway, this question was a little vain, because it was, once again, too mental. For various reasons, the state of bliss that I experienced in the garden or in the woods—not to yet tell of the mountains—could not last. The dusk will arrive at sundown; perhaps a shower of rain will force me to seek shelter, and, obviously, we are not always at the beginning of spring or at the heart of summer.

Fortunately, the desire for a human companion helped me to endure mourning for the impossible permanence of bliss in nature. But I had to internalize what such a bliss meant in order to keep it in memory and wonder about the way of continuing my human journey without forgetting what nature provided me with.

Such an undertaking was really not easy, because my culture did not teach me how to carry it out. Indeed, how could I represent, express, and assess, through words and following Western logic, what I experienced? No doubt, poetry was a possible means to celebrate and praise it, and it was a human manner of expressing my fullness and gratitude—a thing that my tradition does not value as much as would be suitable, and that a publisher put forward as a reason for not publishing Everyday Prayers.

However, if writing poetry brought fullness to me, as staying in nature could do, I still longed for something more. How could I realize what I was longing for?

A first, apparently simple answer was that I longed for reciprocity—initially in touch. What I desired was above all gesture(s), reciprocal gesture(s): to touch and being touched, to embrace and being embraced with someone whose skin was warm, testifying that we were living and could and wanted to share life. Could this gesture be the most basic one in becoming adult(s)?

I acknowledge that, at first, I felt nostalgia for only being embraced by nature, like a newborn or a little child. And, no doubt, this nostalgia sometimes arose again. However, beyond the fact that it will never be completely appeased—nature can wrap me up with the warmth of the sun and the caresses of the wind, or it can surround me with the presence of the vegetal world and of some kind of wild animals, which keep me company, sing to me wonderful airs indeed, but it/she can not hold me in its/her arms—I had grown and also hoped for more mutual relations.

Now such relations required me to become capable of transcendence, including, if not above all, at the level of touch, if I wanted to respect the other as other so as to reach reciprocity. My tradition still ignored this human potential, having left embodied relations at the level of mere instinctive needs or of perverse behaviors, at best, at the service of pro-creation. None of these alternatives allowed me to go further than the bliss that I experienced in nature. To succeed in this, I had to return to and within myself, to assume my human individuation and respect my limits as a separate body. Moreover, I had to seek the way in which my own natural belonging could be, in its turn, a source of felicity for myself and in sharing with the other. To say that in a few words, I had to pass from the infinite that I felt in the communion with nature to an infinite of which I had to take charge as corresponding to my human desire. It is such a quest that encouraged me to venture to return among humans.

11–17 March 2014

9

RISKING TO GO BACK AMONG HUMANS

All is rigid: the ground, the walls, the gestures, the words. All remains fixed, keeps the same form from one day to the next. And we hear the same sounds, smell the same odors or lack of odors, whereas in nature everything changes all the time—a leaf, a flower, a blade of grass, a smell, or a sound changes in one night. However, the forest and the garden give an impression of permanence and safety in spite of constant change. In the city it is quite the opposite: the unchanging state of things is accompanied by a feeling of unsafety. And, instead of moving according to my desire, I had to obey codes, about which I had to inquire.

Not even the day and the night remain. The city is lit from the dusk and the rays of the sun barely come to you, except if you live on a higher floor. It is thus difficult to know where the sun is, or what hour it is. Natural light cannot guide you as it does in the woods, and the smells still less. What about the song of the birds? The pigeons coo the entire day, and the lack of trees or cats mean the absence of other birds. You must learn how to adjust yourself according to the noise of the traffic or the neighbors, the hours of work, the opening of shops, the transportation.

How could I move and grow in such a context? My energy got lost by passing from one codification to another, all being unsuitable for life itself, which consequently lacks bearings and cannot get organized into a world. I lost myself in this constructed order. I turned into a robot. I tried to survive without really living.

But I was there in order to find a companion. How could I direct my research? Sometimes I was so alone that I took the subway to at least feel warm. I saw or met really various people with different behaviors that, from time to time, I found amusing or shocking. I did not know that brushing one's hair, cleaning one's teeth, or doing one's nails in the subway was appropriate. This does not look particularly trendy and does not create a climate favoring intimacy, but, instead, produces a bad distance between people. I also saw people who had a picnic, and that was also more repulsive than appealing. However, most of people were withdrawn into themselves, the ears closed by earphones of some iPod or other. A few were reading and were a little more approachable; sometimes they even gave me a smile.

Generally, the habitués of the subway are led, in this way, to their work or back home. They are only going, and they are barely available for a meeting. It can happen that some are compassionate and even considerate toward you, but going by subway did not allow me to meet a companion. And it was not better on the train. I made acquaintances traveling by train, but this happens less and less because, as soon as they sit, people open their computers. Moreover, this kind of transportation also seems to anesthetize people a little, or to specialize them: they go somewhere and are not free for going toward you. They are with you without really being with: they are transported in the same direction, entrusting themselves to the motion of the machine without initiative or energy of their own. Nothing there resembled what I experienced in nature, neither in the movements of others nor even in mine.

I wondered whether a café would be a better place for meeting someone. But many people were standing at the bar and drank something as speedily as possible, while others were already accompanied or waiting for a meeting. The café did not look like an appropriate place for an encounter either.

Furthermore, in nature all that I met followed a truth that sent me back to my truth, whereas in the city all seemed artificially coded. Each attempted to conform to a truth defined outside themselves, that did not correspond to their own truth, but removed them more and more from a possible return to their self, which could shed light on this truth. Wandering outside themselves, the persons whom I met had no living energy to be shared. Subjected to a cultural elaboration unfaithful to life, each was in exile with regard to oneself, in search of a truth to be respected because it was presumed to be shared by all, though its construction made it shareable only by persons of a certain culture or tradition. Hence a blending of affirmation and uncertainty, of authoritarianism and blind submission to imposed codes, which prevented a sharing not only at the level of judgments but also at the level of sentiments and desires.

It was possible to exchange pieces of information; however, this kind of getting in contact did not allow one to be present to the other, but only to communicate through a third party extraneous to the current encounter—a thing that prevents this from really happening and from being embodied in an appropriate way.

I thus lost not only the vegetal environment, which helped me to keep myself alive and in good physical and mental health, but also the energy potential that sharing with another living being might bring to me. At the level of energy, information, as is generally the case with manufactured objects, takes energy reserves away from ourselves, instead of cultivating or restoring it. Information not only does not provide us with sap for our own growing, but it also uses our vigor to produce or convey forms that substitute themselves for ours. Sharing information sometimes can help us to fit into a given society, culture, or place; however, more often than not, it transforms us into the mechanism of a universe that is not our own. Hence, the fact that we wander, sometimes as an animal that has lost its territory and sometimes as a constructed subject in search of a suitable dwelling place.

Most persons whom I met were in exile with respect to space as well as to time and, above all, in exile with respect to themselves. Therefore, they depended on information, which remained a way for them to become integrated into a society that acted as an artificial familial dwelling place. Maintaining the habits, customs, and existing rules was, thus, the main role of those in charge of the government of the city at all levels, because they substituted for the natural ties uniting the diverse family members.

However, the arbitrarily constructed character of these links prevents relations between persons from happening at a natural level. They are subjected to codifications through which some living energy filters, from time to time, creating a moment of being in communion with one another, a moment that cannot last and be cultivated between the two, because it vanishes into various kinds of encodings. Such a moment could intervene as the lightening that Heidegger speaks of, if we were capable of paying attention to it—what is more, the two at the same time. But this happens exceptionally and can really seldom serve as a common clearing.

Consequently, this lightening provokes misunderstandings, conflicts, and division, whereas it could enlighten each one about oneself and the other at a more global level in comparison with the one Heidegger alludes to and allow for a more complete sharing. Such an enlightenment lies more or less at the origin of the awakening of desire or love, but we pay too little attention to its meaning and its development toward a possible and always more achieved sharing.

Probably, I was in search of this sort of experience, which has something to do with the ray of sun, of light, going through the leaves and the trunks of trees and illuminating the depths of the woods. If Heidegger transposes this phenomenon to the level of thinking, I want to live such an enlightenment with my whole being, and also as a possible conjunction between two inner suns. Are not desire and love a sort of internal sun, the rays of which we long to receive from one another? Could not this happening create a bridge between our natural and our spiritual belongings, and so permit humanity to fulfill itself?

Our tradition did not reveal the importance of such a possible event to us, except, perhaps, when it tried to approach some mystical phenomena linked to the disclosure of a divine being. I think that desire and love between us have always a share in mystical experience and are always confronted with a negative path. How many times must we give up the illusion of a perception of the other as other before getting to be able to meet this other? How many times must we transform our bodily sensations, our feelings, our intuitions and thoughts before approaching a real sharing?

In reality, this cultural sphere remains uncultivated and subjected to rules and norms that do not allow for a sharing between two persons, above all between persons who do not belong to the same world. They remain separated from one another by the modalities in which sensibility is lived or expressed and also by the rest of the instinctual energy that cannot take into account the otherness of the other because it aims at possessing or appropriating more than at sharing. Crossing back over such an erroneous construction of our subjectivity has something in common with the negative path that the mystic must follow to perceive the nature of the divine. Unfortunately, our tradition leaps too quickly over our human condition to get at the divine, and this causes more sufferings than graces, perpetuating the original sin, which deprived us of experiencing happiness in the Garden of Eden.

From then on, we wander in search of ourselves, torn between different parts, first between our body and our spirit, incapable of gathering ourselves and of regaining our wholeness, which cannot occur outside the cultivation of our natural belonging. The absolute we can share is the one that corresponds to the experience of nature as such. Without starting from it, we cannot meet as living beings, because we are each lacerated between diverse cultural constructions that have been substituted for our natural belonging.

I thus met with various embodiments of human being depending on the place in which we were: at work or on holiday, in the street or in public transportation, in a garden or in a café, on a sports ground or a place of prayer, during a conference or in a restaurant . . . In reality, I ran into different roles or functions, but not different people—thus not a possible companion. A culture that is based on information, as ours is, does not consider the education of our entire being enough, but instead turns us into a sort of hard disk that is more or less full, more or less complex and appropriate to one or another culture, tradition, language . . .

However, even if it was henceforth common to have a computer as a companion, this did not correspond to my desire. And contenting myself with a mere instinctual relationship did not suit me anymore either, all the more so since our so-called instincts are already codified in a certain way and no longer express a natural longing. Rather, they show a desperate attempt to free ourselves from cultural norms, which maintain us in exile from our natural belonging. I had recovered a significant part of the latter in the company of the vegetal world, and I did not wish to divide off from it again. Nevertheless, I perceived that clearing a path toward making it blossom into a comprehensive sharing with another human would be a difficult and long undertaking, especially for overcoming what was familiar to us and reaching what was suitable for achieving our humanity and a real sharing of human intimacy.

24 March–6 April, 2014

10

LOSING ONESELF AND ASKING NATURE
FOR HELP AGAIN

Sometimes, meeting humans entertained me, but it also exhausted me. I lost my energy in trying to share. However, I was next sent back home with a show in mind more than a sun shining in my heart and in my whole being—I heard words, saw gestures, remembered some conversations or exchanges, but without the experience of a real presence or of being in communion. In a way, I remained a spectator who was at an encounter in which I could not really participate. And when I attempted to get completely involved in it, the result was quite painful, because I hoped for a follow-up and a development of what happened, but, more often than not, it ended in nothing.

What could I do with the energy that I put into the relationship? Criticizing or lapsing into a feeling of dereliction appeared to me to be lacking interest. I thus continued spending time in nature as often as I could and doing yoga—awaiting, as my yoga teacher advised me. He did not tell me what I had to wait for. At least his recommendation helped me not to continuously lose my energy and to pursue my quest toward a fulfillment of my human being. I also wrote, because I had no choice but to do that—following Rilke's advice to a young poet.[1]

Of course, there were people who considered me egoistic and a social misfit. Fortunately, I had a psychoanalytic training, which allowed me to relativize these criticisms. Furthermore, I did not withdraw from the social milieus that I frequented; I was expelled by them—as I have described in chapter 2. Probably, people wanted me to continue sacrificing myself to a society that transforms us into ghosts or robots so as to never question themselves . . .

One of the most frequent arguments that was put forward was a sort of democratic moralism, at which one cannot even wonder, on pain of being taken for antidemocratic. The weakness of democratic culture too often renders its advocates as much authoritarian as nihilistic. Indeed, assimilating human being to a possessor of money amounts to reducing it to nothing. Money only corresponds to a more or less arbitrary convention that no human decides on. It is tragically laughable to witness the rhetorical efforts some politicians must make today in order to appear to be democratic rulers or leaders. Before pretending to represent other humans, it would be advisable that they wonder about what, or who, is a human being. Alas! they aim to govern humans without first having doubts about humanity itself. Hence, the fact that current democratic regimes are heading straight for disaster. And the most urgent invitation to be given to citizens is that of caring about preserving and cultivating their life, instead of subjecting it to fallacious discourses and illusory promises.

It is the same, in my opinion, with the more and more insistent appeal to respecting the common good. Obviously, this new moralistic slogan is not accompanied by an explanation regarding its content. And, in the same discourse, we may hear an injunction against damaging the so-called common good and an additional attack of the rulers on the modest properties that, step-by-step, we acquired thanks to our daily work. Understanding the meaning of *common* is thus not obvious, except as a last rhetorical device to subject citizens to the ideological plans of the elected, against which we can not even object with the definition of the dictionary. The meaning they attribute to words is that which is required by the most urgent requirements for the exercise of their presupposed democratic power.

No doubt, words lose all signification, or can adopt any signification, if they are not rooted in the real. Now the first common good for which a democratic leader ought to ask for respect is that which concerns life itself and its essential environment. Until democratic programs care primarily about conditions for the life of all citizens, they will remain unconvincing; all the more so as they might deprive citizens of breath, not only at the natural level but also at the cultural level. They are not based on a human, or even not human, living necessity. To what does a government correspond if it does not first care about the preservation and the cultivation of life itself?

They will protest against my interrogation that the family is responsible for life. Can I refer once more to the message presented in the tragedy *Antigone* by Sophocles? To separate the care for life from the government of the city—a thing that Antigone opposes—has serious consequences for both the environment and a culture of life. The rules governing the city become arbitrary with respect to life itself, while the latter falls into a mere natural state again within the family home. The cultural order that Antigone defends against the institution of Creon's power does not meet with such impasses, to which we are facing up in a caricatured way in our time, and which put us in extreme danger. Indeed, we still lack laws relating to the protection of the environment that living beings need and laws regarding the preservation and cultivation of life. What does it mean making up a society in such a case? Is, then, social life anything other than a place for the consumption of life?

Certainly, we have advantages by living in society. However, are these profits not to the detriment of our life? How else could it be in the absence of a culture of life itself? Now, leaping from a mere natural survival to a social membership does not allow for such a cultivation. A stage is lacking. In my opinion, this stage cannot consist of passing from a natural family to a cultural or political family. It requires us to reach our maturity by assuming our sexuate identity as a relational dimension that must act as a natural basis on which any community can be founded.

Indeed, it is our sexual maturity that drives us to leave our original family. The purpose of this estrangement from the family home could be to start a new family. We then miss the stage of our entry into a social milieu as living adults, and the task that is incumbent on us with regard to building a relational world from our natural belonging. This latter remains devoted to begetting and assuming a parental role without a real cultivation of our sexuate identity. In sociocultural environments we behave as sexless individuals whose natural energy is subjected to laws or rules, which are abstract, and often arbitrary, with respect to our living desires.

Obviously, our desires do not first aim to procreate but to create links between us. These links must not confine themselves to so-called private life, or to sexual relations, in the strict sense of the word. They must also serve to weave a social fabric, which corresponds to a cultivation of our relational potential, and does not amount to a construction that is unconcerned about it and based above all on money and goods. A social weaving faithful to our natural identity, and which contributes to its individuation and cultivation, does not exist today. As we neglect to take care of the natural environment essential to our life, we neglect, too, to consider the social environment that we need to develop as humans. How to find a human companion in such a context?

I wondered whether I could receive some help from the reading of philosophers, poets, and from literature in general. But all the authors I approached seemed to come up against the same problem, unless they ignored the question or chose to overlook it. For example, Heidegger, who had a very intense and multifaceted love life, does not reflect in his texts on this relational aspect of human being, in spite of the fact that it represents a crucial path to question and overcome our traditional metaphysics. On this point Hegelian dialectics makes an important contribution to interpreting historically—through referring to the tragedy *Antigone* by Sophocles—the assimilation of sexuate difference to private or public roles, which prevents the woman and the man from meeting and sharing as global beings; however, Hegel does not provide us with a way to resolve the impasses that the duality of complementary but opposite sexuate parts—which have been defined unilaterally by a culture in the masculine—represents. As for Nietzsche, he takes a great interest in feminine figures and characters, especially as representing life; he criticizes their failings with fierceness and affirms that only maternity can correct them; nevertheless, he acknowledges that he needed a feminine companion in order to be able to pursue his work, a companion whom he did not find, and we know the exhaustion of energy that, at least in part, resulted from such a lack.

I am afraid that the philosophers who follow Nietzsche have not perceived the seriousness of his quest for a feminine companion. If Merleau-Ponty, Sartre, and Levinas talk of sexual intercourse in their work, these relations take place in a perspective of domination of nature and not of sharing in or of nature, and this does not correspond to the Nietzschean search for a feminine companion. For Merleau-Ponty and Sartre, the sexual scene has more to do with a master-slave struggle in which sight, as a sense or as consciousness, has the major role; and, for Levinas, touch is still used by the lover to reduce the feminine partner to a presupposed naturality, whereas the man returns to his metaphysical absolute(s). Nothing, thus, in the work of these philosophers answers the tragic appeal of Nietzsche to a feminine companion in order to become capable of crossing the bridge toward a new humanity. The thought of Nietzsche is probably the one that gives rise to the worst misunderstandings, which perhaps explains the terrible crisis of subjectivity and culture that is facing us in the West.

I also read poets and novelists, hoping that I would discover in their works lights to make my way. But I met, above all, with unhappiness, distress, abandonment, and I almost never took delight or received comfort while reading stories or descriptions of beautiful, peaceful, reciprocal, and fecund relations between differently sexuated companions. And, when this happened, it was in a natural environment and with a sharing of love for nature—as in the novels of Susan Fletcher. I thus decided to pursue my quest in the woods or in the mountains.

11–17 April 2014

11

ENCOUNTERING ANOTHER HUMAN IN THE WOODS

For most of my contemporaries, longing to meet a companion in the woods amounts to desiring a reversion to a wild state, to the caveman. And, as people start worrying about how things are going, we begin to see images, spectacles, and discourses concerning primitive man flourish, and interrogations about the evolution of the living arise. For example, to my words regarding the necessity of our returning to nature, a university teacher recently objected, laughing: "You propose to return to Cro-Magnon man!" I answered: "I just wonder about the way to save life." This sort of discussions leaves me completely exhausted, because these persons not only do not acknowledge the energy I bring to them, but they also take my energy, furthermore showing disdain toward me—a situation that is really common in our time.

I do not think that many Western academics know Śiva, the Indian divine figure who corresponds to our era. The main attributes of Śiva are fire and sight; nevertheless, he also has a foot in water and carries a trident, the attribute of the mastery of flows, in one of his hands. Śiva represents the achieved yogi, capable of the greatest concentration. Śiva is also regarded as the god of love. Unlike Kriṣna—and Jesus?—he does not surround himself with a few women, but confines himself to only one. However, two women, or two embodiments of the same one, are in turn his partners. Depending on whether Śiva is in love with one or with the other, he can prove to be more creative or more destructive. The role of the amorous companion of Śiva is thus decisive for the destiny of the world and of all living beings. Parvati, called the white partner of Śiva, is the one who is able to keep the fire going so that it can be creative. Parvati is also called the woman of the mountains, that is, of the place where the air is more subtle and fresh, and so can help breath to ascend and become spiritual. In reality, the figure of Parvati has something in common with that of Mary, who also appears in the hills and embodies the fresh and moderate part of the Spirit in the Christian tradition. Kali, for her side, is called the black woman of Śiva, and, instead of soothing the passion of the god, she exacerbates it so much that the fire often becomes only destructive.

Śiva can be viewed as the god who is in charge of the passage from one age to another and who can carry out this task according to whether his companion is Parvati or Kali. Śiva appears to me as the god who could help to assure the transition between the Nietzschean old man of the West and the new man, imprudently called "superman" by Nietzsche. It is also possible, in my opinion, to compare the present era of Śiva to a third time of Judeo-Christianity, the time of the Spirit, which is also dominated by fire. Unfortunately, the Western tradition does not consider the role of sexuate difference enough in the accomplishment of the passage from one era to the other. And this explains why the inspired intuition of Nietzsche about his need for a feminine companion to pursue his work remained unrecognized. May I also quote on this occasion the meaningful words of the novelist Almudena Grandes in *Inés and the Joy*: "immortal history is often a love story."[1]

Śiva is the god of the transformation of living being and of love. He does not belong to the same culture and does not date from the same era as Nietzsche. Nevertheless, this divine figure partly offers a solution to the crisis that faced Nietzsche and with which to confront our epoch.

Śiva overtly embodies in himself the four elements that form the macro- and the microcosm, and the place in which they gather and change into one another. One of his feet remains plunged into water, while the other is already dancing, showing the capacity of the god to surmount the earth's gravity; one of his hands holds fire and the other a trident that represents domination over the rivers. All that is present and in transformation in Śiva himself, so much so that contemplating him we contemplate the universal metabolism of the world.

We imagine with difficulty Nietzsche ensuring the passage from the old man of the West to the new man acting as Śiva, even if he suggests that it is the way allowing for such a transition. However, Nietzsche, in spite of the attention he pays to the body, too often confines himself, for example, to allusions to, metaphors or transpositions of the dance instead of really dancing. But the effect on the transformation of matter is not the same! No doubt, Nietzsche takes a more important interest in the state of the body than most of the other Western philosophers, and the words of Heidegger on this aspect of Nietzsche's thought are not always correct. But it is true that Nietzsche has not discovered the manner of cultivating his physical energy, and his interest in the body can often look more psychological than physiological. This bears witness to our traditional negligence, and even forgetting, of a cultivation of our natural belonging. Nietzsche also mistakes reflection on past forms of our culture for the emergence of new forms, something that doing yoga and meditation can bring, or at least he favors too much the former gesture over the latter.

As an accomplished yogi, Śiva practices meditation, and this corresponds to a concentration and a spiritual transformation of his energy, indeed not only to a thinking about something. For want of a reserve and a cultivation of his natural energy, Nietzsche could not pursue his undertaking or even his own life. Nietzsche made a mistake, too, regarding the place in which to dwell. He chose the mountains, and even the ice, and not the forest, which is Śiva's favorite place for dwelling.

To favor a high and fresh place made sense for Nietzsche, but the risk also existed of repeating the freeze of thought as it happened in the Western tradition, or that of not going further than an interpretation of the latter by remaining situated in an outside point, allowing for a historical perspective on it. In my opinion, it is the altitude at which Nietzsche was standing that has contributed to providing him with an intuition about the eternal return of the same. The perspective brought by the difference of level where he was, as well as the change in the density of the air, and thus of breath that went with altitude, has permitted Nietzsche to leave the circle within which he was kept captive by his tradition.

However, this did not suffice to disclose to him how to become the superman—or the new man—he wanted to become. For such a happening, the woods were preferable. And it is not a matter of chance that only the hermit that Nietzsche met in the forest had not yet heard about the death of God.[2] Perhaps, instead of being already dead, the god that we could meet in the forest is still to come. The forest is a place where escaping the opposition between freezing and burning is possible. As it is possible there to avoid the distinction between matter and form—*hylē* and *morphē*—which is at the root of our cultural tradition. The woods are an environment where matter produces its own forms without being reduced to a material to which we, as humans, have to give forms. The woods show us what being living consists of. Beyond the fact that they provide us with the material elements we need to survive and grow as living beings, the woods teach us that, as long as it remains alive, matter—*hylē*—produces its own form(s)—*morphē*. When we pretend we are those in charge of giving form(s) to matter, we act as denigrators of life. Furthermore, we, then, substitute a demiurge power for the task of producing our own forms as living.

In a way, Śiva is the Hindu god who shows us how to behave in the woods—he stays in the forest in order to reach his individuation, especially through meditating, dancing, and loving. He dwells in the vegetal world in order to achieve his destiny: to become a living being capable of producing his own forms. He does not coexist with plants in order to become plant, to imitate or appropriate plant being, but to be capable of becoming who he is—to allude to a Nietzschean teaching.

In order not to be satisfied with a passive communion with the vegetal world, Śiva must actively cultivate his own energy and lead it to an appropriate embodiment. Meditation and dance contribute toward the concentration and transformation of living energy. Śiva also needs a suitable companion for awakening his masculine energy, for preserving it from freezing or burning so that it does not become sterile or destructive, and for keeping living energy available for a future becoming.

21–25 April 2014

12

WONDERING HOW TO CULTIVATE
OUR LIVING ENERGY

The Western tradition has neglected the cultivation of natural energy as such. It favored human fabrication over natural growth and production. Even if early Greek culture still focused on the contemplation of what nature gives us to see, from the Greek golden age onward, this natural appearing began to be depreciated and supplanted by human plans and action regarding nature.

During its first intervention in the Sophocles tragedy *Antigone*, the chorus, which represents the voice of the people, warns Creon of the danger of substituting man's deeds for an economy faithful to nature. But Creon does not listen to the popular wisdom, and he sentences Antigone to death because she defends natural values against a masculine power that aims at dominating them.

The natural rights that Antigone tries to protect, at the risk of her life, include three aspects that are linked together: the respect for the cosmic order as an environment necessary for all living beings and under god's tutelage; the respect for natural generation over human fabrication; the respect for sexuate difference as a dimension of our bodily identity, which preserves us from being reduced both to a mere anonymous matter after our death and to a neutralized undifferentiated individual during our life. Our difference in relation to the other(s) then results only from the role we assume in society and from the quantity of our material or spiritual goods. This difference is already cut off from our natural belonging and prevents us from keeping and cultivating it. The articulation between our natural identity and our cultural productions is lacking, which renders the fulfillment of our humanity impossible. Furthermore, the energy that is used to produce a presumed human world is a neutralized one, an energy that is removed from living beings and transformed into an abstract strength in the service of more or less despotic rulers, who impose their power through arbitrary laws.

There is no place there for a cultivation of our natural energy, and when it arises, especially when it is awakened by sexual desire, it must be discharged as a potential for our own growth and subjected to a process of reproduction of future citizens. Unless it is molded in accordance with the norms already effective in the city. In reality, we have to sacrifice our personal energy and growth to a common order, which has become parallel to the order of the living. Consequently, we do not know what to do with our living energy but expend it, investing it in a sociocultural economy that dispossesses us of it or suspending it on suprasensitive values that prevent us from embodying it. Anyway, this energy does not remain available for our own growing and for living relations to and with the other(s) and the world.

If Heidegger has deep and beautiful words about the work of the hand, including at the level of thinking,[1] he remains silent about the modeling of our own being, although this probably leads the way toward a new horizon of thought. He was well aware of the fact that such an opening up was necessary, but he could only prepare it without yet perceiving its nature—to allude to his own words. It is true that it is no longer a question of a work of our hand alone, but of a process that concerns our complete being and calls for a more radical change in our manner of thinking and behaving.

Such an evolution requires us to question the four causes that intervene in the fabrication of a work, according to Aristotle: the material cause, the formal cause, the efficient cause, and the final cause. In the work that we have to accomplish toward reaching our own being, these four causes interact in a way different from that analyzed by Aristotle. Indeed, we are ourselves together matter, form, maker, and end. We are no longer making something, be it a spiritual product as thought. We are working for us becoming ourselves—following the Nietzschean teaching "Become who you are!"—an undertaking that is not as simple as making a cup, to quote the famous example used to explain the four Aristotelian causes. Now, the matter is living, and it already has form(s) and aim(s); the form can be shaped, but not ex nihilo, and not only according to our own intention; we are both making and made and we cannot anticipate and decide on the final destination of the work.

We are entering into another economy of thinking and of living, in which the Heideggerian split between being and Being no longer operates, because they are altering one another all the time, without one representing the presumed essence in relation to the other. Ultimately, it is rather our being that becomes the essence from which we have to grow. Obviously, the word *essence* then assumes another sense: it means an original potential from which we must start and develop. But this potential is never pure, as a traditional essence is. Beyond the fact that we were born from a man and a woman, that is, that our chromosomes and genes are already hybridized, we are also conceived in a cultural context, which acted upon those who have generated us and the first environment in which we began to live. Contrary to the vegetal world, our "essence" is never simple, and this renders our human roots more complex and uncertain. Hence, the fact that we need the vegetal as a shelter and a help for our life.

The constant interaction between our being and Being—I would prefer to say "being" and "to be"—questions the horizon within which the Heideggerian and more generally the Western way of thinking develops and, probably, brings some answers to the quest of the philosopher.

It also modifies our conception of time itself, allowing us to surmount resentment against the past, which, according to Nietzsche, is the fundamental reason that prevents us from leaving the old man for a new humanity.[2] Henceforth there is no longer Being as such, and there is no longer time as such. We are, or ought to be, all the time working being and time so that we can only reproach ourselves for not taking charge of becoming who we are, without feeling resentful toward time itself or whatever reality seemingly external or extraneous to ourselves instead of toward the way we, ourselves, conceived our path. In fact, our resentment arises from our dissatisfaction because we have not worked on a becoming of our humanity.

We are not able to carry out the passage to a new human being without returning to our natural belonging, recovering the living energy it provides us with and learning how to use it in another way. So living in harmony with the seasonal rhythm helps us to go back to our natural roots and to maintain our vigor more than conforming to a constructed temporality, and we would be well advised to shape our existence in accordance with the annual cycle. For example, we must take care not to blindly spend the additional strength given to us by the spring, but to think of the growth to which we devote to it, and we must not ask ourselves that we try harder during the winter, resorting to various medicines in order to succeed in this, but, instead, return within ourselves, gather ourselves to keep alive and prepare the seeds of a future becoming.

This aspect of the cultivation of our energy is not the most difficult to implement and could even, at least partly, be in harmony with our Western tradition—as the texts of some philosophers and writers show. But we lack directions regarding the energy awakened by sexual desire. It is probably why our tradition only tells us of amorous misfortunes, when it does not consider sexual attraction to be a lapse or a sin. It can speak of amorous awakening, but not of human fulfillment and sharing of carnal love. Even most of the Eastern traditions neglect this dimension of our humanity, crucial for its achievement. For example, if Confucianism treats the question of a culture of natural ties, it confines its discourse to relations toward the parents or other members of the family, without taking the cultivation of sexual love into consideration for our becoming humans. The cultivation of our sexuate belonging and sharing seems an undertaking that we have still to approach in order to discover an economy of our living energy.

26 April–4 May 2014

13

COULD GESTURES AND WORDS SUBSTITUTE FOR THE ELEMENTS?

Perhaps cultivating our sexuate belonging corresponds to making a way toward becoming humans. Indeed, this will contribute to our human individuation. The vegetal world is not sexuate as we are, and the animal world does not use sexuation as we do, or ought to do. The growth of plants is determined by an interaction between the elements that constitute the living universe, without an effective connection between different plants of the same kind. If plants have a soul, this seems to remain neuter, and their reproduction does not happen through sexual attraction and relation between two plants, but thanks to a third that conveys the necessary germ or seed—be this third the wind or an insect—and thanks to a cycle of reproduction in which the soil acts as a nurturing receptacle. Anyway, a plant does not have to divert its growing in order to reproduce. Even if this may need lateral ramifications, these do not divide from the roots or from a vertical development that, above all, obeys solar attraction.

In reality, Platonic thought in part imitates the vegetal way of growing, whereas it renders it impossible: the good—the Good—as Plato defines it, cannot substitute for the sun. What is more, humans do not have an origin as simple and unique as plants, beyond the fact that they move, among other things to fulfill their desire, especially their sexual desire. Plato did not worry about these essential properties of the human being, which makes his way of conceiving development inappropriate to us. But we are still trapped in the Platonic horizon, and turning upside down the Platonic model of growing does not suffice for going out of it. We must add to the Platonic perspective some elements that it has neglected, though they are crucial for human becoming.

In an oversimplified manner, I could say that if the relation to sexuation of the vegetal kingdom does not require it to move away from original roots and give up a unique and vertical growing, and if the animal kingdom seems to favor a horizontal way of moving in order to satisfy needs and instincts, notably sexual instincts, humans must develop by combining a vertical and a horizontal growing, which makes their becoming more complex and largely still to come. Human beings remain torn between a vertical and a horizontal accomplishment without having yet reached their own way of becoming. Not only are their roots never one but at least two, but their growing also intertwines with that of others, especially others who are different; which renders their becoming hybrid and uncertain in its motion and direction.

There is no doubt that our language as it is, our *logos* and its evolution, has not allowed for a complete blossoming of our humanity. Our culture still lacks the means to cultivate life itself and the crucial property of the living that sexuation is.

At the level of life as such, perhaps we ought to care about a language that not only alludes to the elements, their potential and their intervention in the generation and the maintenance of life, but also, in some way, acts as them. Could we discover words and a manner of speaking capable of providing us with air, water, fire, and even earth, instead of constituting a discourse that appropriates elemental beings without keeping their living qualities? Could we succeed in shaping a language that is not only parallel to the living world but is itself living and participates in cultivating life? For example, could our way of using words bring to the other breath, warmth, humidity, and terrestrial belonging and dwelling, or must we also invent new words? Unless we have to associate both operations?

And how to mix gestures and words with such an undertaking in mind? I foresee that we could adopt a manner of relating to one another that, without substituting for the elements, would intervene one way or another as them. This corresponds to experiences that I live but cannot yet express clearly. If I try to communicate something of them, I could say that when someone smiles at me I feel both warmth and light, as the sun can bring to me. And it is probably not by chance that, in their drawings, children often represent the sun smiling. If someone touches me, I am brought back to my bodily and terrestrial existence, with sometimes an additional aquatic perception, particularly if I am touched through words. If someone inspires me with admiration or enthusiasm, be these linked to his or her person or an eventual plan, it is breath that is solicited from me, and sometimes also my heart. Unfortunately, most of the time our manner of behaving and speaking neglects these aspects of a cultivation of life, and we enter into relations as robots coded by a linguistic program rather than as living beings.

Our language even deprives us of being fully receptive to what the elements can bring to the maintenance and the growth of our life. Our mind has been transformed into a sort of hard disk that has taken in that which our education has imposed on us, without becoming a center capable of gathering and governing all the components of our living being. Our culture as it is has paralyzed a possible evolution of our global being, especially regarding its relational properties. So much so that we can no longer enter into relations with one another as living, and the living energy we could, then, develop and share no longer exists. We can only reproduce at a merely natural level, but our cultural and spiritual fertility has been sterilized by our culture. How can we rectify such an error, and undertake from now on the development of our living potential?

This potential has remained uncultivated, especially with regard to the energy resources of our sexual desire. Now, it is above all this energy, as a living one, which can act as our most human sap. Indeed, its development and its cultivation can provide us not only with a natural growing—in a way similar to plant growing—but with a specifically human embodiment. If our physical growth is dependent on the intervention and the interaction of the elements that compose the living world, and if this growth is already endowed with forms, as a plant is, it is incumbent upon us to further shape our natural belonging in a human way. The question is not one of imposing fabricated forms on our natural forms, but of cultivating them—I could also say: of using them, with the meaning that Heidegger gives to the word *use* in the second part, section 7, of his book *What Is Called Thinking?*—with our human becoming in mind.[1] A task that sends us back to the usage of the four Aristotelian causes in the work of our becoming humans.

How to use the energy of our sexuate desire in this task? It seems that our tradition has imagined our desire was similar either to vegetal sap or to animal instinct, that is, that it amounted to a vital strength, a strength that aims to rise higher and higher toward the sun—or the Good—or to move laterally to possess its "object," especially for reproducing. Yet our human desire has other possibilities that we must realize in order to accomplish our human destiny.

No doubt it is not an easy undertaking, notably because desire does not fall into our linguistic and logical categories: it does not merely exist by itself; it is not simply visible, not even easily perceptible; it does not obey only a vertical or a horizontal direction; it does not have definitive forms, but can adopt and give form(s). This latter potential is probably the most crucial for our human growth. Indeed, desire is able to create new forms in comparison with those with which we are provided by nature. However, desire must respect these forms to contribute to a cultivation of our humanity. Our current education and culture still lack this perspective, essential for our human, especially our relational, becoming. Desire is not awakened only by our own nature, even if fertilized by the elements, but by an other, especially another human. This additional and, originally, natural energy arises from a source that does not lie simply in ourselves and requires us to take into account the link with the other in order to develop, to be embodied, and to be shared.

We cannot remain only passively receptive to desire, particularly to amorous desire. Perhaps, we ought to live it as an internal sun that allows us to grow in a specifically human way. Indeed, we have not to transform the light and warmth of this sun into an immediate bodily, and even spiritual, spreading or manifestation, but we must actively let them inhabit and modify ourselves until they build an intimate dwelling that helps us gather ourselves, to return to and within ourselves so that we are able to welcome the presence of the other without either appropriating it, uprooting this other from a living existence, or losing ourselves in a fusional process or a blind submission. Cultivating desire asks us to remain two, which requires us to use and develop our natural forms in a human way, without obeying the immediate impetus awakened by an external cause, especially by the other.

Such an evolution necessitates our respect for the elements composing the macrocosm and their appropriation to the microcosm that we are. Appropriating, here, does not mean entering into possession, but making suitable for a specific usage. Some traditions—the yoga tradition, for example—know and cultivate the possible transformation of our bodily and elemental belonging toward its human development. Even in our religious tradition(s) we find some allusions to a transubstantiation or a transfiguration of our physical nature. However we still lack teaching about how to make the way toward such an evolution regarding our relational, especially our sexuate, accomplishment. It is a perspective, or a horizon, that I try to approach and elaborate in some of my books, such as *I Love to You, To Be Two, The Way of Love, In the Beginning, She Was,* and *A New Culture of Energy* (this last still to be published in English).[2]

2–9 July 2014

14

FROM BEING ALONE IN NATURE
TO BEING TWO IN LOVE

Are we alone in nature? Obviously not completely. We are surrounded by many living beings with which we can be in communication, even in communion. And, more often than not, this happens more easily than with humans and provides us with an experience of the absolute that we attain with greater difficulty with a human.

I might recall here examples of lovers who became nature lovers because they did not succeed in reaching a communion with the person with whom they were in love. We would not have the admirable poems of Hölderlin if this were not the case.

I might also wonder, on this occasion, about Hegel's renouncement of an experience of the absolute in sharing with nature instead of working on the transformation of this experience into a spiritual absolute. What results from this separation of one absolute from the other? Our separation from life itself? Making a way that has been entrusted only to a mental awareness and a visual intuition, leaving behind a cultivation of our embodied being? Could this correspond to the end of a human journey? And does this not amount to cutting off the end from the beginning, contrary to the assertion of Hegel himself? How could we, then, pass from a communion with the vegetal world to a communion with another human different from ourselves? We are now deprived of a living energy that could be used as a medium for making this communion possible. We are separated from an experience of life itself and of our living environment and, henceforth, we approach them only through constructed codes that keep us in exile from our natural and living belonging.

Then we are alone in nature because our relation to life has become objectified, notably by naming and integrating all beings into a whole, so that we can no longer be in communion with them. Furthermore, through our way of constructing the world, we have separated all living beings from one another and deprived them of the link(s) between them, that is, of a source of life, which has more to do with breath and touch than with sight and whatever description and reproduction. Hence, our world looks like a sort of museum composed of inanimate things invested with our projections. Because of our want to organize all of the real according to our intentions or wishes, we are left on our own in the midst of the world that we experience as lifeless and treat as such. Which harms each living being, including ourselves, now compared to fabricated objects instead of remaining enlivened by our own sap, our own vigor, and those resulting from our exchanges.

Most of the time, we have lost the perception of the difference between meeting living beings and meeting things, be they material or spiritual, that we made. Only when we are confronted with the resistance of the other, we pay attention to its existence. Unfortunately, this implies that we mistake the living for an obstacle we must master, without enjoying being in communion with it. It is so much so that we have defined a lot of rules in order to settle our relations with all instead of entrusting the meetings to our longing for sharing and cultivating life together. This aspiration, when it remains, is thus paralyzed by codes or moral imperatives that break up any living link.

Nevertheless, it happens that life is stronger than the conventions and orderliness that prevent it from existing and manifesting itself. It is the case when desire, especially sexuate desire, oversteps the limits of the imposed codes and customs. Obviously, the needs also testify to the existence of life. But, more often than not, they cause competition, subjection, or even destruction, and do not favor a harmonious communion between all the living. Sexuate desire is, at least originally, an appeal for entering into relations with the other as a source and an embodiment of life different from ours, which calls for a sort of becoming characteristic of us as humans. Indeed, we are then forced to enter into presence not only through appearing, but as a being present to the other. We are not developing starting only from our own roots and in a single environment; we have to compromise with other roots and another world— a constraint that requires us to stop only growing in order to wonder about the existence of the other. And if desire compels us to pause for a moment to consider the presence of the other, it can so contribute to awakening our attention to and also respect for any living being, before which we generally forget to linger and wonder.

The desire for another human can revive an interest in life that we often neglect. Desire calls us back to life, but we barely listen to this call, and we submit our longing too to norms and rules that are unconcerned about life. This happens, above all, when we leap over the relationships between two in order to enter into a community or a society. Desire, then, loses its living roots, unless it falls back into more or less wild instincts. It is bent by cultural constructions that do not act in favor of a sharing of life.

It is between two, and two humans different by nature, that a living sharing can be found again. The face-to-face with a human different from us gives us a chance of leaving the constructed world within which we are trapped and of recovering our global being. We need to meet a "you," and a "you" different by nature, to be able to regain our natural belonging and to start from it. For lack of such an encounter, we get bogged down in a maze of constructions that deprive us of the experience and sharing of life, and we remain fragmented in many parts and investments that we are not even capable of gathering together.

Sometimes we recover a unity through having faith in a religious or philosophical Absolute. However, at least in our tradition, this generally cuts us off from sexuate and mutual attractions and sharings, instead of teaching us how to develop them in a human, and even a divine, way. And yet such a cultivation of our sexuate surges is crucial for our becoming able to behave as a living being among other different living beings without domination or subjection. It is when our energetic impetuses are not used as they ought to be used that we lose the perception of the limits of life as such, and so we become destructive, including toward ourselves. Unfortunately, the ideal values, which hold our sexuate desires in abeyance, notably by their pretense of being universal and in the neuter without contributing to their embodiment and fulfillment, bring about our oscillations between good and evil, right and wrong, because of the lack of a suitable consideration for our energy.

Intertwining desire and love is certainly a better manner of clearing our path, beginning with the place where sexuate desire arises within us and between us. Transforming the source and the additional living energy that sexuate, and especially sexual, attraction represents into an amorous union is probably the way toward the achievement of our human being and the acquisition of a human manner of behaving and being in communion with all living beings.

It is not by chance that, in some traditions, a couple of gods in love is represented as being at the origin of the world or of a new age of its evolution. Did our tradition make the mistake of entrusting this original foundation to a couple of humans? Or have we not yet reached the maturity that would allow us to assume such a task?

23–29 July 2014

15

BECOMING HUMANS

Perhaps things are this way because, in our tradition, we have imagined that the original couple of humans was not engendered and was, then, worrying about becoming god before accomplishing their humanity. They were in search of their divine origin instead of becoming the humans who they were. All the more so since they were presumed to govern a world already created and did not know how to carry out this task. They lacked the perspective from which they could emerge from the whole in order to value everything and decide on it.

Hence, their want of appropriating knowledge of good and evil like children who feel the fear of being disliked by their parents. Unless what they willed to possess was the aptitude for distinguishing between right and wrong without going through their own experience. Which still amounts to taking over for their own use the knowledge of the other—of the Other?—without making the way for their own knowing. Their knowledge is a kind of mastery cut off from a living experience.

I am afraid that the greatest part of our culture is founded on such an appropriation. We, then, consider knowledge to be a good that we must acquire from the past more than a way of being at present. And culture is an instrument of domination over the real more than a wisdom that allows us to coexist with everyone and everything through cultivating the part of the real that we are. Being cultivated thus separates us from all, including from ourselves, instead of rendering us capable of developing who we are and contributing toward the development of each living being. Now this aim would amount to governing the world with justice, and it requires love from us—a way of loving that is not only impetus, sentimentalism, or morality, but is respectful evaluation in our consideration for everyone and everything and in our behavior toward them.

What we still lack is to join loving and thinking: a becoming human that we cannot realize as long as we remain children dependent on whatever family authority. The perpetuation of such a status from the most natural level to the most spiritual level keeps us both irresponsible and arrogant. Furthermore, it allocates to us imitation as the path of growing. Becoming human adults would imply that we have become some "monkeys," as one says, capable of appropriating knowledge of their sire as a tool for their survival.

However, what we, as humans, must appropriate is what is useful for our integration into a society that does not care a lot about our life—our social rules and conventions are based on the neutralization of the living more than on its respect and its cultivation. Our education teaches us to move away from life until we lose the way to return to it, in ourselves and in every living being that we meet. We learn how to behave in a certain sociocultural context, but not how to deal with the living world—which would allow us to become living beings among other living beings and not mechanisms somehow or other codified.

Therefore, the misunderstanding and conflicts between different cultures or traditions and their respective claims to represent the universal truth that must be imposed on all. Only life is universal, and starting from life we can build a human culture and accomplish our humanity. This asks us to reconsider our sociocultural foundations, especially the language we use to enter into relationships between us, but also with our living environment.

For example, denomination is probably our most basic cultural process. But how to assign a name to what is through becoming and changing, as is the case for life? Furthermore, if some cultures have words to express the motion of being, our Western culture lacks them. In Japanese we could probably say that the tree trees, that is, is becoming tree through its growing, as it is possible to say that the cloud clouds.[1] We do not have this opportunity. For us, Westerners, a tree can be an oak, a pine, a beech, etc., and we learn how to designate and classify the kinds of trees without taking a real interest in their living properties and needs. We act toward vegetal beings as toward inanimate things, outside any consideration for living relations with them and between them.

It is almost the same regarding animals and humans. And if our language is capable of passing a judgment on them—which amounts to a sort of classification—it does not provide us with the terms and syntactic structures that favor communication between us. Our linguistic tool is elaborated for mastering more than for cultivating and sharing life. It is so much so that we come to designate as a so-called real what is already the product of our viewpoint on it, not even "our" viewpoint but that which has been taught to us.

In fact, we go round in circles more and more remote from life and living exchanges. Our language is more and more coded, and the technical means we employ to express ourselves and communicate from a distance make it gradually weaker and dead. Some people try to revive it through abbreviations or diminutives, but, that way, language loses sense again in spite of its seductive familiarity and apparent convenience.

Rather, the question is to rethink our language in order that it expresses life, ours and that of other living beings. No doubt, favoring sight in the approach to all has not contributed toward a culture of life, as did not putting the stress on names, and not on verbs and adjectives.

The surplus neurons with which humans are provided unfortunately have been used either for developing life with love and intelligence or for creating a human existence starting from the natural life received by birth. They have been used for dominating what maintains life and its growth: the sap, the vigor and energy that animates the breath, the blood, especially when we are in love and in communion with another living being and with the living world.

We have neglected to consider such a resource and have constructed a culture that thwarts this potential, instead of letting it be and making it blossom, which has rendered the living world sterile, except regarding reproduction. Now, life can be fertile for many things other than reproducing. We have not developed and made flower our living belonging, and our surplus neurons; our "head" has more often than not acted against and not in favor of our embodiment. This has led to the depreciation of the woman, the one who gives birth with her blood to a being made of flesh and blood. Is not our education based on the scorn for flesh, blood, even blood ties, and the attempt to master them through laws, rules, and discourses arbitrary with respect to life? Which results, at best, in ambivalence toward all that has something to do with living, in us and around us.

Our bloodless sight then decides on being through its appearing and not its hidden existence and growth. Hence the perpetual search for the cause and the origin outside of living beings while forgetting their own. Hence, also, the fact that humanity has defined itself in opposition to other species, especially the vegetal and animal species that are more simply related to life, and has underestimated the importance of the difference between the genera for the efflorescence of life, and not only for its reproduction. In order to differentiate itself from other living beings, humanity had then to consider itself to be one and capable of dominating the living world instead of being a living being among others, more capable of sharing with all thanks to its additional awareness and freedom.

3–7 August 2014

16

CULTIVATING AND SHARING
LIFE BETWEEN ALL

More and more, our manner of behaving, and even of thinking, is based on appearance(s) to the detriment of living. This exhausts our vital reserves and those of the earth without supplying us with new resources. The future of the whole living world is henceforth endangered. And there is no doubt that we must urgently worry about such a peril and be concerned with building culture and our systems of exchanges on new foundations. We must start from life again as the only value that can be universally shareable and learn how to cultivate it with the preservation and the blossoming of all beings in mind. We must thus focus on what we have in common and the way of safeguarding this common good.

The elements—air, water, fire, earth—are the most basic thing about which we must care. Even before their specific intervention in the formation of each being, they must be taken into account as the condition for the existence of life and its necessary environment. I am afraid that most people have never thought of their composition from the four elements and of the fact that harming them amounts to harming themselves. They ignore their material constitution and do not take care of it, especially as the real through which they partake in the living world. If they were to pay more attention to that, they would become aware of the necessity of caring about their environment as much as about their own life, because they are made of the same material.

I am really surprised that the doctors do not inform their patients of such a fact, and that a relay of various techniques, in both discourses and medicines, prevents then from becoming aware of this basic reality, of which our ancestors and the living beings of other species are more conscious than we are. Nevertheless, if the sciences and their products can prolong our survival, they do not contribute to a growth and a cultivation of life as such. In most cases, they maintain people in a state of dependence on a knowledge that acts as a parental and authoritarian power, which deprives them of the freedom that they need for becoming. Now, nature brings us autonomy from birth through breathing, and it continues providing us with the means of developing and maintaining our life by remaining receptive to the elements it offers us. Besides, this way, we stay in a grateful communion with the whole living world.

However, we are humans, and if the elements are necessary for our existence, we must also reach our own individuation, which requires us to give up being continuously in communion with them. Nature abhors a vacuum, but we are forced, as humans, to take it on, without renouncing our natural belonging. It is so difficult to join the two that this unsolved task has hindered or perverted our human becoming. Even Hegel did not succeed in solving the question, but it is true that his work consisted in summing up the story of Western philosophy more than inaugurating a new epoch of thought. He thus tried to overcome the negative as well as our natural identity instead of discovering a logic which makes their compatibility and articulation possible. Now, this challenge is specifically human, and it is worth grappling with in order to both ensure a proper becoming and allow for suitable relations with the other as other, especially the other different from us by nature. Indeed, we cannot, as humans, simply obey the impetus awakened by an external cause. We can no longer celebrate the presence of the sun, which arouses our desire, with new leaves or flowers as the plants do, with songs, as the birds do, or with an impulsive making love, even if all that is already a manner of praising, of which humans are too often careless. We must also cultivate and mold what we receive with the achievement of our human destiny in coexistence with other living beings in mind. And this demands of us being able to suspend a mere bodily growth and to keep available a part of the wakened energy for a transformation that does not amount to an immediate and appearing growth, but is necessary for meeting and sharing with respect for the otherness of any other. In fact, we have to convert our own material belonging into another physical matter, without subjecting it to abstract forms and norms extraneous to it. We must defer our immediate growing in order to shape and pursue our becoming in accordance with our human properties and potential. And this gesture requires us to be confronted with a first form of the void in relation a natural continuity and contiguity.

I do not think, as Hegel does, that the question is, then, to overcome the immediacy of what is experienced through its negation by our awareness, a negation that itself will be finally surmounted in reaching the Absolute. Rather, I think that we have to take on the void and preserve it as an insurmountable moment of our becoming. Our way of growing cannot be merely continuous: its development needs a discontinuity that permits our own individuation and respectful relations to and with other living beings. This is another reason why the Platonic, and more generally our past metaphysical models are inappropriate to the cultivation of our nature and are basically nihilistic. Instead of assuming the void as a condition of our human individuation and development, they pretend to mark out the way to expel the void and the negative from the world thanks to ideals that are out of our reach and impose, on the real, constructed patterns that prevent communication and communion between beings as living. The paradox regarding living beings is that they must keep separate in order to share. They cannot share without being faithful to their own sap, their own natural resources of blood and vigor.

However, desire is another way of entering into relations that can transcend the ties of blood while being faithful to our natural belonging. Desire, and first sexuate desire, corresponds to another level of our individuation that does not amount to a mere material and vital belonging. Desire has to do with a beyond in comparison with mere survival. It is a force that compels us to overcome the subjection to biological needs toward reaching a human and spiritual accomplishment. But desire is already dependent on forms—originally, on physical and bodily forms—that shape our identity and our manner of dwelling on earth. Desire can transform a simple territory into a world.

Desire is basically sexuate, which does not mean necessarily sexual. Its contents and forms are neither neuter nor indifferent to our bodily constitution, but they do not always aim at sexual intercourse. Which does not prevent our relations—to the world, to the other(s), and to ourselves—from being sexuate. Our tradition has misjudged the importance of this fact, a mistake that has harmed our human achievement and our relational way of behaving with all living beings. Once more, the void, the nothing, and the negative have been suspended on, or supposedly annihilated by, suprasensitive ideals that have harnessed our energy without allowing for its growing and its sharing between us.

In reality, and for the second time, we have to take on the void, the nothing, the negative existing between us as naturally different in order to be able to share sexuate desire. And this gesture ought to intervene before any sexual relations, so opening a place where they can happen as a fulfillment of a human desire and not the satisfaction of presumed sexual needs.

The question regarding sexual needs, especially masculine sexual needs, that have to be satisfied at any cost, results, in my opinion, from a lack of the cultivation of our sexuate belonging toward an embodied fulfillment respectful of the other who is different from us by nature. Then the energy aroused by sexual attraction remains without means to shape up and express itself, and it asks for a discharge or a release out of consideration for the other involved in the relation.

What is at the origin of our relational life as desire has not been taken into account as a crucial aspect and condition of our human development. Such an omission has distorted our manner of relating to other humans and to the whole living world. It has also perverted our perception of the real, our valuation of truth and of the way to reach and share it. Cultivating life toward a sharing between all requires us to care not only about our vital preservation and growth but also about our relational potential, beginning with molding the energetic resources resulting from our sexuate belonging in all our relations.

Sexuate energy must be neither kept for sexual relations in a so-called private context, where they largely remain uncultured and above all devoted to reproduction, nor ignored and repressed in a sociocultural environment where we ought to behave as sexless individuals. Sexuate energy represents an important living source and reserve for our own becoming and our taking a place among other living beings.

Taking on the void or the nothing of nature in growing, which requires our becoming as humans, and taking on the void or the nothing in relating to and with the other(s), which requires respect for otherness, already presuppose an aptitude for thinking. Before any articulate language, humanity is, or ought to be, characterized by such an aptitude. Thinking is not the privilege of some philosophers; it partakes in being human, and it calls for an ability to distance oneself from an experience that is only immediate, a necessity that is generally mistaken for mastering the real through objectivizing it. We ought almost to gesture in an inverse way, that is, momentarily consider ourselves objectively in order to decide how to pursue our journey, starting from a more correct and living perception of the real, instead of projecting on it the sociocultural constructions we have made ours.[1]

Thinking asks for taking on the void or the nothing, for a third time, as a renouncement of adherence to our sociocultural background or to the knowledge that is already ours. Thinking calls for being confronted with the void or the emptiness without aiming at only overcoming them. Indeed, such an experience provides us with a perspective that does not necessitate separation from the living real.

Perhaps Western tradition lacks a cultivation of breathing, which allows for facing and dealing with emptiness without dying or falling into nothingness. Breathing is crucial for living, loving, and thinking, and its economy, as the nature of breath, varies depending on whatever and how we embody our existence. From birth, we must breathe by ourselves. Sharing with the living world, with the loved one, or with human or not human others cannot happen without cultivating our breathing. For lack of caring about it, we pervert life, ours and that of all with whom we supposedly share it, because we are not able to respect, to love, and to think of each in their living otherness and give them back to their own roots and growth.

By breathing, we also open and reopen continuously the possibility of a new growth and horizon for life, for desire, for love, and for culture. Sometimes the vegetal world is our most crucial mediator; sometimes it is a loving and loved human different from us; sometimes past thinkers lead us on the way. However, we must clear up our own path alone, with the help of a star, of a grace, or of the fecundity of a meeting with another human who longs to cultivate life, love, and thought through their sharing, and with building a new world in mind.

9–19 August 2014

EPILOGUE

*plants are
key to
all survival*

Dear Michael,

When I suggested that you co-author this book I did not imagine how this would happen. As I mentioned in the letter that introduces the volume, two main things have prompted me to make such a proposal. The first is the fact that our earth and all the living beings who inhabit on it are henceforth in danger and that the preservation of the vegetal world is crucial for trying to save the planet. The second is that we, as humans, must find a way to escape putting ourselves at the disposal of the domination of techniques and the sciences, which today rule the world, through a return to our natural belonging and its suitable cultivation for the establishment of another manner of existing and coexisting that allows us to keep our energy alive.

Personally, I think that sexuate identity is the framework—I could say the *Gestell*, as and with a slightly different meaning from Heidegger—which makes possible such an undertaking. In fact our sexuate identity is what permits us to transcend the mere materiality of our body toward its transformation into a specific cultural individuation and a relational order based on the growth of life and its sharing. It was thus not indifferent from this point of view that I imagined a man could be a better co-author of a book that invites us to go back to a cultivation of life, especially with regard for the vegetal world and the teachings it can provide us with.

No doubt, some people will immediately think of sexuality on this occasion, but this results from a lack of culture of life and its sharing that makes our sexuate identity a mere opportunity to waste our living energy on. Obviously, my wish was quite different. It was to attempt to reach a dialogue in difference through the vegetal world. Hence, my proposal to choose the title *through* rather than *with* vegetal being. Then *through,* for me, means that we live thanks to vegetal being, which procures pure air for us; that vegetal being somehow corresponds to a stage of our own becoming; and that it is also the environment through which we can leave our past tradition, in which we were divided into neutralized body and soul, toward a world that takes into account our identity and subjectivity as they are, that is, as sexuated.

However, calling into question and trying to leave the sociocultural background and context that have molded not only our behaviors and our manner of thinking but also those of our human, and even not human, surroundings does not occur without endangering our existence. The help of the vegetal world is thus crucial for carrying out such an undertaking without risking dying or falling into a nihilism worse than the one that undermines our tradition.

Furthermore, the vegetal world is the one that can allow us to discover what it means to be and to act in faithfulness to our sexuation without reducing this specific identity to traditional sexual relations, instead of wondering about its role in conceiving what a living truth could be. Indeed how could we ethically deal with the vegetal world if we ourselves do not remain living at all levels of our existence, including in thinking? There is, or ought to be, a sort of dialectical process between caring about vegetal being and being faithful to our sexuate identity and subjectivity. One cannot occur without the other, and both require us to overcome our past metaphysical method of reasoning and behaving toward a new way of thinking, talking, and sharing.

This cannot happen overnight! And if readers imagine that they will find in this volume a traditional dialogue about vegetal being, they might be surprised. But if they believe that they will discover here what a dialogue in difference could be, they are also mistaken. Rather, they will look at a counterpoint between two perspectives, something that, hopefully, could pave the way toward a future dialogue between two different manners of thinking and of saying. Indeed, such an opportunity is a preliminary event, and it has already called for behaviors respectful of our mutual differences. If we did not yet reach the words and the syntax that allowed for a dialogue between us, as a man and a woman, at least we prepare its possibility through endeavoring to respect the alterity of the other, to listen to a different thought, to leave space and time for the manifestation of otherness.Such gestures prove, among others, a capability for keeping silence and standing back in order to let the other appear and enter into presence. This corresponds to the first imperative to meet any living being—be it vegetal, animal, or human—and form a living world with it. It is through the respect for the transcendence of living being that we can open up and build a world that does not lack transcendence without being subjected to suprasensitive values, which do not favor the blooming of the living as such. In reality, consideration for the vegetal world cannot go on while remaining within a logic of the same, and the Same, but requires us to adopt a logic found in the respect for difference(s) between living beings, beginning with us as differently sexuated.

The path will be long until we understand that we have considered to be most unworthy of cultivation the part of our natural belonging that is at the very origin of our longing for transcendence!

I thank you very much for having ventured to take a few steps with me toward the unveiling of the place where a possible *Ereignis* of a becoming of humanity, crucial for the safety of the world, could still happen.

I want to give thanks to the readers too. They were present in this undertaking and they have contributed to transforming a personal journey into the attempt to elaborate a phenomenology of a global human being, especially a feminine one, that is still lacking. I hope that from my collaboration with you many will receive some light for clearing the way for a cultivation of life and for their own growth toward the construction of a possible and better future for all.

Luce Irigaray

Written or rewritten between 24 September 2014 and the beginning of 2015

NOTES

PROLOGUE

1. Gianni Vattimo and Michael Marder, eds., *Deconstructing Zionism: A Critique of Political Metaphysics* (New York: Bloomsbury, 2013).
2. Michael Marder, *Plant-Thinking: A Philosophy of Vegetal Life* (New York: Columbia University Press, 2013).
3. Michael Marder, *The Philosopher's Plant: An Intellectual Herbarium* (New York: Columbia University Press, 2014).
4. Luce Irigaray, *Speculum of the Other Woman*, trans. Gillian C. Gill (Ithaca, NY: Cornell University Press, 1985).
5. Cf. Friedrich Hölderlin, *The Poems of Friedrich Hölderlin*, trans. James Mitchell (San Francisco: Ithuriel's Spear, 2004), pp. 7–15.
6. Luce Irigaray, *Sharing the World* (New York: Continuum, 2008).

1. SEEKING REFUGE IN THE VEGETAL WORLD

1. Michael Marder, *The Philosopher's Plant: An Intellectual Herbarium* (New York: Columbia University Press, 2014).
2. Martin Heidegger, "The Origin of the Work of Art," in *Basic Writings*, ed. David Farrell Krell, rev. ed. (New York: Harper and Row, 1993), pp. 139–212.

2. A CULTURE FORGETFUL OF LIFE

1. Luce Irigaray, "Animal Compassion," in *Animal Philosophy*, ed. Matthew Calarco and Peter Atterton (New York: Continuum, 2004), pp. 195–201.
2. Luce Irigaray, *To Be Two*, trans. Monique Rhodes and Marco Cocito-Monoc (New York: Continuum, 2000).
3. Luce Irigaray, *In the Beginning, She Was* (New York: Bloomsbury, 2013), p. 115.

3. SHARING UNIVERSAL BREATHING

1. Luce Irigaray, *Between East and West* (New York: Columbia University Press, 2003), and *A New Culture of Energy and the Mystery of Mary* (New York: Columbia University Press, forthcoming in 2016).
2. Mary Lutyens, *Krishnamurti: The Years of Awakening* (Boston: Shambhala, 1997).
3. Edwin Bryant, ed., *The Yoga Sutras of Patañjali* (New York: North Point, 2009); *The Upanishads*, trans. Juan Mascaro (New York: Penguin, 1965); *The Bhagavad Gita*, trans. Laura L. Patton (New York: Penguin, 2008); *The Yoga of Spiritual Devotion: A Modern Translation of the Narada Bhakti Sutras*, trans. Prem Prakash (Rochester: Inner Traditions, 1998).

4. THE GENERATIVE POTENTIAL OF THE ELEMENTS

1. Cf. Luce Irigaray, *Marine Lover of Friedrich Nietzsche*, trans. Gillian C. Gill (New York: Columbia University Press, 1991).
2. See Luce Irigaray, *The Forgetting of Air: In Martin Heidegger*, trans. Mary Beth Mader (Austin: University of Texas Press, 1999).

5. LIVING AT THE RHYTHM OF THE SEASONS

1. See, for example, Heinrich Zimmer, *The King and the Corpse: Tales of the Soul's Conquest of Evil* (Bollingen Foundation, 1957).
2. Cf. Samuel Noah Kramer, *The Sacred Marriage Rite: Aspects of Faith, Myth, and Ritual in Ancient Sumer* (Bloomington: Indiana University Press, 1970).
3. Aeschylus, *The Complete Aeschylus*, vol. 1: *The Oresteia,* ed. Peter Burian and Alan Shapiro (Oxford: Oxford University Press, 2011), pp. 30–31n27 (translation modified).
4. Susan Fletcher, *Witch Light* (London: Fourth Estate, 2011), p. 45.

6. A RECOVERY OF THE AMAZING DIVERSITY OF NATURAL PRESENCE

1. See the first gathering of these poems in the bilingual edition of my *Everyday Prayers*, trans. Luce Irigaray, with Timothy Matthews (Nottingham: University of Nottingham and Paris: Maisonneuve and Larose, 2004).
2. Cf. Luce Irigaray, *Being Two, How Many Eyes Have We?*, trans. Luce Irigaray, with Catherine Busson, Jim Mooney, Heidi Bostic, and Stephen Pluhacek (Rüsselsheim: Christel Göttert, 2000).

7. CULTIVATING OUR SENSORY PERCEPTIONS

1. Cf. Luce Irigaray, *Being Two, How Many Eyes Have We?*, trans. Luce Irigaray, with Catherine Busson, Jim Mooney, Heidi Bostic, and Stephen Pluhacek (Rüsselsheim: Christel Göttert, 2000), taken up again in "Dialogues," a special issue of *Paragraph* 25 (2) (Edinburgh University Press, 2002).
2. See the section on " 'Being-with' (Mitsein) and the 'We,' " in Jean-Paul Sartre's *Being and Nothingness*, trans. Hazel Barnes (New York: Citadel, 2001).
3. Patanjali, *Yoga Sutras* (New York: North Point, 2009).

8. FEELING NOSTALGIA FOR A HUMAN COMPANION

1. Luce Irigaray, *Everyday Prayers*, trans. Luce Irigaray, with Timothy Matthews (Nottingham: University of Nottingham and Paris: Maisonneuve and Larose, 2004).
2. See ibid. as an example of poems written between August 1997 and July 1998, 116.
3. Friedrich Nietzsche, *The Gay Science*, trans. Walter Kaufmann (New York: Vintage, 1974), fragment 125.
4. Cf. Luce Irigaray, *I Love to You: Sketch for a Felicity Within History*, trans. Alison Martin (New York: Routledge, 1996), chapters 9, 10; *To Be Two*, trans. Monique Rhodes and Marco Cocito-Monoc (New York: Continuum, 2000), chapters 2, 10; *The Way of Love*, trans. Heidi Bostic and Steven Pluhacek (New York: Continuum, 2002), chapters 2, 3, 4; *Sharing the World* (New York: Continuum, 2008), introduction and chapters 3, 4.

10. LOSING ONESELF AND ASKING NATURE
FOR HELP AGAIN

1. See Rainer Maria Rilke, *Letters to a Young Poet*, trans. Charlie Louth (New York: Penguin, 2014).

11. ENCOUNTERING ANOTHER HUMAN IN THE WOODS

1. "La Historia inmortal es, a menudo, una historia de amor." Almudena Grandes, *Inés y la Alegría* (Barcelona: Maxi Tusquets, 2011), p. 28.
2. Cf. Friedrich Nietzsche, *Thus Spoke Zarathustra*, trans. Adrian Del Caro (Cambridge: Cambridge University Press, 2006), section 6, part 4.

12. WONDERING HOW TO CULTIVATE OUR
LIVING ENERGY

1. Cf. Martin Heidegger, *What Is Called Thinking?* trans. J. Glenn Gray (New York: Harper Perennial, 1976).
2. Cf. Friedrich Nietzsche, *Thus Spoke Zarathustra*, trans. Adrian Del Caro (Cambridge: Cambridge University Press, 2006), especially part 2, section 7.

13. COULD GESTURES AND WORDS SUBSTITUTE
FOR THE ELEMENTS?

1. Martin Heidegger, *What Is Called Thinking?* trans. J. Glenn Gray (New York: Harper Perennial, 1976).
2. Luce Irigaray, *I Love to You*, trans. Alison Martin (New York: Routledge, 1996); *To Be Two*, trans. Monique Rhodes and Marco F. Cocito Monoc (New York: Continuum, 2000); *The Way of Love*, trans. Heidi Bostic and Stephen Pluhacek (New York: Continuum, 2002); *In the Beginning, She Was* (New Yrok: Bloomsbury, 2013); *A New Culture of Energy* (New York: Columbia University Press, forthcoming).

15. BECOMING HUMANS

1. Cf. a dialogue between Heidegger and a Japanese master in Martin Heidegger, *On the Way to Language,* trans. Peter D. Hertz (San Francisco: HarperCollins, 1982), pp. 1–54.

16. CULTIVATING AND SHARING LIFE BETWEEN ALL

1. Cf. Luce Irigaray, "Pour une logique de l'intersubjectivité dans la différence," XXV Internationaler Hegel Kongress: Das Leben Denken, *Hegel Jahrbuch* (2007): 325.

MICHAEL MARDER

PROLOGUE

Dear Luce Irigaray,

In *Sharing the World* you write of "the risk to open one's world in order to move forward to meet with another world."[1] When I first read these lines, still before encountering you through our correspondence, I was unaware of the extent to which you followed this imperative in your daily living practice. In our exchanges and in what I have come to learn about your mentoring of graduate students, your generosity has been and remains unmatched to that of any other philosopher I know. You are not afraid to risk opening your world to others with a candor and energy I can only admire. But, to share the world, it would be necessary to share the risk too—without seeking protection behind the heavy armor of criticism. It is in this spirit that our communication has proceeded thus far and it is in this spirit that I take the perceptive questions you have posed in your most recent letter.

Before venturing anything like a response, which, in any event, will not be exhaustive but will require further work of thinking, I would like to recall one of the intentions behind our dialogue that may flower into a book. A return to nature, especially to vegetal nature, is impossible outside the cultivation of humanity as a relation—a sharing of the world or of worlds—at least between two. Just as the elaborations of the meaning of being human are deficient and one-sided when they do not account for sexuate difference, so our relation to the nonhuman world is stunted if it does not develop with the shared contribution of all differently sexuated human beings. My question for you in this regard is: How many worlds participate in this relation? Yours and mine, to be sure, as well as, perhaps, our shared world. But what about plants? Do they, too, have or constitute a world? Can they be "others"? Or do they belong somewhere on the hither side of the distinction between the same and the other?

In *Plant-Thinking*,[2] my wager (a considerably risky one at that) was to try and restore to plants their world, which is in excess of a "physical environment." That is why, for me, the question is: How can our worlds (together, in common) encounter and share with the world of plants? Whether this kind of a formulation makes sense or not, however, depends on how, if at all, we articulate plants with the categories of traditional metaphysics, including sameness and otherness.

In light of this concern, the vegetal deconstruction of metaphysics I undertook was much more than an intellectual exercise. An expanded ethics that could grow to encompass our treatment of plants and, consequently, the very uncertain future of life and of care for life was at stake in it. This critique of metaphysics from the standpoint of plant-being, which has been so demeaned and abused along the intellectual history of the West, would not have been worthwhile if it were to stop at a simple inversion of values—for instance, privileging multiplicity and plurality over unity. I tried to explain this in my chapter "The Plant That Is Not One," included in a collection edited by Artemy Magun and titled *Politics of the One*,[3] which revolves around the thought of Jean-Luc Nancy. The one plant, I wrote there, is not one—a multiplicity (of growths) that does not merely negate the one, or the One, but reassembles it in a community of growing beings. Far from a negation of universality, plants, as growing beings par excellence, are, in my view, the figures of singular universality, to which differently sexuated human beings also belong. (After all, before it appeared in the animal kingdom, sexuality was actually a part of plant life.) I am likewise wholly sympathetic to your experience of having rediscovered in the vegetal world some aspects of Greek culture, which still permitted its subjects to sense the proximity of the whole of nature—*phusis*—to the plant—*phuton*, that is to say, of an ensemble of growths to a "growing thing." Again, here the plant reemerges as a singular universal that oversteps the bounds of the metaphysical tradition and relates to all of nature.

Probably the most testing aspect of *Plant-Thinking* is that it constitutes neither a simple continuation of this tradition nor an absolute rupture with it. My readings thus far have been situated at the margins of metaphysics, showing or disclosing its internal limits, above all, when this tradition aims to process plants through its indifferent conceptual machinery. It seems to me, nonetheless, that I have also begun to affirm something other than a variation on the deconstruction of metaphysics. As far as the time, freedom, and wisdom of plants are concerned, plant-thinking offers an alternative vision of the being of plants, of which we partake either consciously or, as is most often the case, unconsciously. Rather than deepening or otherwise elaborating on our nihilistic heritage, the values of "vegetal existentiality" point beyond nihilism and its much-celebrated obverse side.

But what about the language that could help us approximate the world of plants—if it is, strictly speaking, a world?—you ask. I agree that this language is still largely absent and that one of our greatest challenges is to assist it in coming into being, to see to it that it could attain its full expression, without, at the same time, violating the silence of plants. Nothing less than a paradigm shift in our current idea of discursivity would do here.

To begin with, not all language unfolds in words; gestures and living bodies speak as well, even in the way they inhabit places. I have hinted in my previous work at the spatial self-expression of plants. Theorists of performance studies have since then adopted this suggestion, broadening the scope of their engagements to vegetation. In a different way, Benjamin invokes in his writings what he calls "the language of things," while Heidegger speaks of the "totality-of-significations," none of which relies on ready-made words or signs. Even silent "showing without saying," seemingly pertaining to the teaching of the vegetal world, fits Husserl's concept of indication as opposed to expression.

More important, I would be reluctant to conclude that only human beings speak while the rest is pure noise or deafening silence. We have not yet learned how to listen to others—and not only to our human others at that. Learning to listen does not mean that we would be able to obtain "more information," previously withheld from us. Nor does it imply that whatever or whoever we would be capable of listening to would become fully transparent and decipherable. The proliferation of multiple meanings, deriving from different worlds, is dependent on achieving a balance between listening and noninterference, learning from the other or from others and respecting her, his, its, or their incommunicable otherness.

And so—to address your third question—the dilemma of passing onto others the experience one has with and of the vegetal world appears to be a part of a more general aporia of sharing. As Derrida would have said, only that which cannot be shared is worth sharing. Is this not the case with the experience of intimacy with the vegetal world? Does not the risk of opening one's world—and its secret allegiance to that of plants— to the other ineluctably entail the risk of losing oneself, along with that which was to be shared? I believe that *that* is the very heart of risk, the riskiness of risk. The point here would be that, at times forgetting myself, I would be able to find the other and, thanks to this discovery, cultivate a richer relation to the vegetal world. The difficult path we have chosen demands—not sacrifice or self-sacrifice, no!, but—a joint relation to the world of plants and, through it, to nature, within the experience of sexuate difference. And, if that is so, then our individual experiences of intimacy with plants are already, to a greater or lesser extent, "betrayals" of what a human approach to plants will have been like, unaccomplishable by either of us alone.

Besides the question regarding the world and the otherness of plants, I would like to ask you about something else, indeed about an issue that has already surfaced in our correspondence. How do you propose to think about the sexuality of plants? What is its relevance to human sexuate difference? What I have in mind is the fluidity, pliability, and plasticity of vegetal sexuality, where many plants are hermaphrodites, others can change from masculine to feminine (and vice versa) in their lifetimes, while still others reproduce asexually. Without a doubt, sexuate difference belongs to the phenomenon of embodiment and to life itself. But is it such a linear process, whereby one is recalled to (or one recalls) one's own sexuate being, as a living body, upon contact with plants? Does the vegetal world open sexuate difference, through which we attempt to encounter it, to sexuate differences that are more diverse still than what Freud referred to as the "polymorphous perversity" of the human infant?

With this series of questions that have been on my mind for some time now, I close the letter, while looking forward to the next chapter in our shared thinking and work. With each other and with plants . . .

My warmest regards and enthusiasm,

Michael Marder

Lisbon, 11–18 December 2013

1

SEEKING REFUGE IN
THE VEGETAL WORLD

The plants' rootedness in a place, their fidelity to the soil, is something we can only admire, especially because our condition is that of an increasing and merciless uprooting. If politicians across Europe celebrate "mobility," which is often the least preferable option for the vast population of unemployed people forced to move to another country in search of a job, it is because the opposite (staying in place, abiding, dwelling abidingly) is associated with stagnation, an incapacity to adapt to the changing circumstances, and a sentimental relic of the past. Of our more vegetal past, perhaps?

Even before immigrating for the first time with my family at the age of thirteen, I felt an acute sense of nonbelonging and displacement as a Jewish child growing up in the capital of Russia. Onboard the plane heading to Israel/Palestine—the land that, more so than all the others, will have never been nor become "mine"—I could not have imagined how many other dislocations would await me: another immigration to Canada, the move to New York for my doctoral studies, a brief return to Canada and other parts of the U.S., the departure to Portugal and the Basque Country in northern Spain . . . In each of these places, plants have become the keepsakes of my memories, the mnemonic centers of gravity that evoke the events and even the atmosphere of my life at the time, down to minute details. A tall birch tree that was planted by my grandfather and grew opposite the window of my room in Moscow; a palm tree and an overgrown cactus inhabiting the backyard in Israel/Palestine; small strawberry bushes and tulip bulbs I brought as a gift from Holland and planted with my mother in the town of Maple, Canada; a young olive tree, commemorating a person who has been unconditionally welcoming toward me in Portugal—these are but a few examples of the vegetal points of repose in my life. This almost automatic habit of associating places with plants (that, unlike me, have remained rooted in their environments) is, likely, behind my reimagining of the history of philosophy in *The Philosopher's Plant*,[1] a book in which the landscape of thought was associated with flowers, trees, and grasses.

Whether purposefully or not, I have been weaving the fabric of meaning starting with plants, though not of one piece with them. Allotting to the vegetal world a place close to the center of psychic life gave me a certain sense of security; the plants stayed, while I left. That is why it was particularly sad for me to realize that, in the mayhem of yet another move from Maple, Ontario, my mother had displaced the tulip bulbs, dug out so as to entrust them to a family friend. For once, not only humans but also plants were uprooted—but not transplanted. And I felt that something within me, too, was uprooted forever, without a chance of recovery, perhaps more so than in all the other displacements.

Of course, the matrix of vegetal meaning changes along with us. When I came back to Moscow, for the first time after a seventeen-year absence, I was struck by the relative smallness of the birch tree growing opposite the window I could now observe only from the outside. What for a child seemed like a gigantic tree was a common birch, though still clothed in layers of personal and cultural significance that made it much more than a tree. (I cannot help but think of Sergey Yesenin's poem "The White Birch Underneath My Window," which we had to learn by heart at school.) Is seeking refuge in the vegetal world tantamount not only to being in nature, with plants, but also to a reorganization of our lives, our routines, our psychic space around them? After all, the world of plants sheltered me physically, and above all mentally, by providing me with a comforting sense of continuity, which was completely lacking in my own life.

I took solace in the fact that, even though, due to racism, militarism, or a mutually reinforcing toxic mix of both, I was virtually expelled from the first two countries where I resided, the plants that grew there continued to thrive. Until, that is, I learned about the uprooting of olive groves in Palestine by the Israeli army. Besides the strong associations of olive trees or branches with peace, made all but unattainable by such behavior, for me, these despicable actions were equivalent to the excoriation of the center of meaning, the semantic core I located in plants. Accompanied by the destruction of Palestinian houses, the uprooting of trees reinforced a transformation of the entire population—the entire people—into refugees, prevented from seeking meaning and refuge (even) in the vegetal world. To my eyes these actions, already decried in ancient laws that intuited in the uprooting of trees the declaration of a total war without a chance for reconciliation, completed the destruction of Israeli legitimacy. Although I was not subject to the kind of violence experienced by the Palestinian people, my subsequent displacement from that small corner of the Middle East was attributable to the same cause.

On the outskirts of Moscow, the place where I was born and grew up was situated at the edge of a massive forest called *Losinyi Ostrov* ("The Island of the Moose") and recognized as the first national park in Russia. The forest was a source of entertainment all year round; there we enjoyed frequent walks and, depending on the season, berry gathering or cross-country skiing. I came to know its pathways, running close to my apartment block, already in my early childhood and felt at home in it. Despite its extension, the forest did not appear threatening, but was, literally, a place of refuge from the heavy pollution, noise, and overcrowding of the city, not to mention of crammed communal apartments.

It was not until my arrival in Canada that I rediscovered the wonders of the forest, strikingly similar to the one I had known from my childhood. But it was this similarity that made more vivid the fact that it was not *the same* forest, which made the memory of loss more acute. Without wishing to generalize this experience, it seems to me that such a memory cannot but haunt our relation to plants. Seeking refuge in the vegetal world, we return to it as refugees: too uprooted from what we call nature, or, more concretely, from our most intimate growth, which we share with everything and everyone that lives. We escape into the plant world, from which we have been fleeing for millennia now. And it is tremendously difficult to undo the effects of this flight that has come to define our culture itself. Nothing less will do than reversing the direction of culture and reconceiving it as a loving cultivation of the vegetal world and of the living.

Having said that, plants provide us with a very peculiar shelter where the traditional distinction between interiority and exteriority no longer applies. For too long, our psychic and physical dwelling places have been constructed in such a way as to separate us from the threatening outside world, until we cut ourselves off almost completely and lost touch with it. Those of us who have been expelled from our social environments, communities, or countries, however, sensed on our very skin an additional separation from the dwelling. That is, perhaps, why we have sought refuge in the plant world. But this world does not provide us with a mere substitute for the lost human dwelling; upon returning to it, we do not recoil into yet another interiority. On the contrary, we find refuge in the absolute exposure—of and to plants, the elements, or a new sort of energy, following the model of vegetal growth. Isn't the condition of a refugee, a reject, or an outcast propitious to a return to plant nature? And aren't the more or less secure psychic, cultural, and physical modes of dwelling, by the same token, the modes of our expulsion from this nature?

Portuguese pine forests and the beech forests covering the mountains of the Basque Country are my new reality: welcoming and already familiar, albeit a little melancholy. I have learned to recognize myself in them, to draw meaning and to organize my world around them. More than anywhere else, the pines growing on the shores of the Atlantic Ocean along the Portuguese coast are capable of gathering together the diverse elements of the sky, the immense body of water stretching all the way to the Americas, and the sandy earth, wherein they are rooted. Like the temple in Heidegger's "The Origin of the Work of Art,"[2] the pines (and, at bottom, all plants) not only *are* but also create a world; they *world*, in the verbal, active sense of the word. That is where I seek refuge, which does not take the shape of a closed dwelling but leads toward the edge of another world.

17–21 December 2013

2

A CULTURE FORGETFUL OF LIFE

The controversy began almost one year prior to the publication of *Plant-Thinking*. On 28 April 2012, the *New York Times* featured my opinion piece titled "If Peas Can Talk, Should We Eat Them?" in which I discussed the ethical implications of recent research on the capacities of pea plants to communicate to one another through biochemical substances, released into the soil by the roots. There, I envisioned the ethics of caring for vegetal life that would "not dictate how to treat the specimen of *Pisum sativum*, or any other plant," for that matter, but would, rather, "urge us to respond, each time anew, to the question of how, in thinking and eating, to say 'yes' to plants."[1]

During the weeks and months that followed, my argument was attacked by a wide range of dogmatic opponents, from Christian fundamentalists to vegans and from neuroscientists to humanist rationalists. The heated polemics on the place of plants in our culture turned personal, when I received various pieces of correspondence containing suggestions to commit suicide, wishes for my speedy demise, and the like. The anonymous nature of contemporary communication allowed some readers to spill out their unconscious fantasies and desires, for which the dividing line between someone advocating on behalf of vegetal life and that life itself was quite blurred. In retrospect, it has become evident to me that a small portion of the deadly and nihilistic energy that our culture tends to unleash against life ricocheted in my direction. An affirmation of vegetal existence in its own right was much more than this culture could bear because a long time ago—indeed, at its very *roots*—it swapped a caring cultivation for the productive destruction of plants.

What brought certain scientists, philosophers, and religious fanatics together was their forgetting of diverse lives in favor of Life, whether conceived as otherworldly existence, or as an objectively decipherable DNA code, or, again, as the purified and sterile "life of the mind." (The case of vegans was slightly different: they succumbed to a narrow defense of one kind of life, that of animals, at the expense of other kinds of vitality.) It is this forgetting that stands at the source of the grotesque, upside-down world we inhabit, where life-affirming philosophy is perceived as the height of nihilism. A world where concern for plants passes as a lack of care for animals and humans. One where age-old prejudices about the meaning of life and intelligence wear the mask of "common sense" and are counted among the outcomes of "clear, rational thinking."

When *Plant-Thinking* was finally released in the beginning of 2013, the reception of the book was highly polarized. Some readers thought that it heralded a new era in our relation to nonhuman life, while others went so far as to raise suspicions that the book was an elaborate hoax.[2] What exactly is a hoax? In and of itself, the word is rather telling. Via a not-so-veiled reference to a magical *hocus-pocus* (mock Latin), it harkens back to the key phrase of the Catholic Eucharist, *hoc est corpus meum*— "Here is my body." The liturgical transubstantiation of bread into the body of Christ is the etymological source of the hoax, in which someone (say, an author) announces about a body of writing or another kind of work that it is her or his body, while, in fact, it is not. Such disingenuity is especially dangerous to the bourgeois social and political order, in which the first institution of property necessarily passes through the appropriation of the body: in the first instance, my own and, gradually, that of nature, with which I mix the efforts of my labor. A fraudulent *hoc est corpus meum* thus threatens the entire system of the artificial production and reproduction of life that depends on this first, deadening act of appropriation, itself the biggest fraud the world has ever known.

Why were some convinced that *Plant-Thinking* was a hoax? Precisely because the conversion of everything and everyone, including various living beings, into objects to be possessed has donned the appearance of an unquestioned norm. It seemed ridiculous to assert that plants were the subjects of their world, rather than the more or less impassive growing *things*, the role they have been allotted within the tradition of Western metaphysics. Moreover, it seemed strange that recent achievements in European philosophy could be "mapped" onto the ontology of plants, presumably far removed from such haughty notions as freedom or intelligence. Without giving it a second thought, the proponents of the hoax theory assumed that, if it was not an outright prank, the book executed a magical sleight of hand by anthropomorphizing plants or ascribing to them properly human characteristics. The core argument of *Plant-Thinking*—namely, that humans share intelligence and life with plants, from which much of our own thinking and being derive—remained obscured by their superficial reading.

Of course, the real tricksters are the firm believers in the received wisdom of metaphysics. Their signature hoax, responsible for today's predominant worldview, is the act of transubstantiating all forms of life into lifeless property. The ideally disembodied, presumably gender-neutral (though actually masculine) protagonist of metaphysics has not ceased to pronounce his fraudulent *hoc est corpus meum* upon encountering anything and anyone in his environs. Kant's distinction between appropriable things and appropriating persons did little to silence this magical incantation of "dispassionate reason." On the contrary, Kant strengthened the total logic of appropriation by simply sorting beings into its subjects and objects, not to mention by consenting to the inclusion of animals and plants in the category of appropriable things.

The true mystery of the Eucharist lies in the vegetal constitution of the body of god and of the human. To say, about fermented wheat, "this is my body" and, about fermented grapes, "this is my blood" is not to transform them into one's property, let alone into a purely spiritual shadow of nature, but to discover or rediscover oneself in parts of the plant world. *Plant-Thinking* sought the path of this discovery through philosophy rid of its anthropocentric bias, rather than through religion. If many of my fellow philosophers were unwilling to give thoughtful consideration to this alternative, then at least a few of the vanguard scientists and artists were receptive toward the way of relating to and perceiving the vitality of plants I had only begun to sketch out.

Insofar as the book was concerned, its *hoc est corpus meum* was, at times, interpreted as a hoax, meaning that this body of writing itself was deemed to be somehow insincere. That accusation was supposed to function as a barrier preventing my argument from entering the realm of what could be debated in respectable philosophical circles and, more significantly for me, as a prohibition against stating that my body of writing is truly my "own." But what would it mean to stand by one's text? Would it entail coming out on the digital *agora* of cyberspace with the public disclaimer "For me, this book is the truth, the whole truth, and nothing but the truth"? Or are there more complex issues concerning the ethics of appropriation—above all of bodies, be they living and breathing, corpses, or corpuses—in the hoax accusation?

In a culture built on property relations, withholding the possibility of appropriating one's own body and whatever surrounds it is the most basic form of economic and social oppression. The woman, the worker, the animal, and, certainly, the plant are construed as either incapable or undeserving of self-possession and, therefore, as fit for appropriation by others. Since Aristotle, this list also includes all those who do not follow the ground rules of formal logic and whose bodies of writing and thought are therefore little more than a hoax. In the case of Antigone, Creon disallowed her to claim both her own living body and the corpse of her brother, Polynices. Antigone simply could not say *hoc est corpus meum* about herself or about the body of a close family member she wanted to bury. We would be amiss, however, if we assumed that everything could be resolved peacefully and culminate in a happy end if only Antigone were able to appropriate, as a sovereign and autonomous subject, herself and the remains of her dead brother. This is the point the classical interpretation, stressing a symmetrical conflict between human and divine law, inevitably overlooks.

Faced with the corpse of her brother, Antigone does not wish to claim it for herself, that is to say, for her family, tradition, or even in the name of the law she abides by, the law Creon considers to be a hoax. Rather, she wants to dispense it back to the elements and, in particular, to the element that is most appropriate to it, the earth. Much more than a simple desire to restore the "natural order of things," her care for the dead is care for life itself. For instance, by burying Polynices, she aspires to free the air for the living, which is exactly what Creon denies them, insofar as he locks Antigone up in the cave and lets a putrefying body pollute the atmosphere. He uses the very smell of death and the sight of decomposition as the insignia of his power. She asks nothing for herself; *hoc est corpus meum* is not a gesture of appropriation for her. She claims her brother's corpse in order to pass it on to the larger world around her. We can only surmise, since no one but Antigone herself could confirm or deny this hypothesis, that her self-appropriation, the cultivation of her subjectivity, follows the same route of finding herself without disrespecting the elemental realm to which we all belong. (Need I say that *Plant-Thinking* did not elaborate a model for the appropriation of the vegetal world, but only aimed to restitute the body of the plant back to the soul of plants, as chapter 1 of that book makes clear?)

In turn, Creon's order, his law and his word, are nothing but a hoax. He endeavors to take hold of bodies that do not rightfully belong to him, bodies both living and dead, and, through them, to control organic and inorganic nature as such. His pretense is that these bodies are only moved or not moved (as the case may be) by the grace of his own murderous will, which he has surreptitiously substituted for life's plural directionalities. The same applies to our contemporary nihilism, which perpetuates this hoax on a global, if not on a cosmic, scale. To maintain its power, it must be in a position to convince us that the living are actually dead unless endowed with a sufficient dose of negation and self-negation. On this view, plants are the least alive of creatures because they do not contain as much as a trace of self-negation, will, or subjectivity; in a word, they do not belong to themselves. Completely delivered to external appropriation, their vegetal bodies only become animated when they are chopped down, plucked, or culled. On the other hand, those who take the side of this "unmediated" plant life are seen as naive simpletons or, conversely, as the cunning satirists who intend to make a joke out of the bearishly serious status quo.

In the more pessimistic moments, it appears to me that a dialogue between the nihilistic culture forgetful of life and the culture of caring for and cultivating life is impossible. The truth of the latter is a dangerous hocus-pocus in the eyes of the former. A veritable abyss extends between Creon and Antigone. How to dispel this forgetfulness? Simple reminders wouldn't do here, seeing that contemporary nihilism only makes itself still more rigid and closes off the public sphere of debate as soon as alternative views are sounded. When dialogue fails, however, pure violence ensues. Survival becomes, more so than ever before, a matter of struggle, in the course of which some will give up (or will be forced to give up) and at the end of which all might lose because what is now endangered is not just the existence of an isolated human being but the prospects for the survival of life itself.

Instead of sending persistent but fruitless reminders to a culture forgetful of life, we must cultivate a different culture, starting from a drastically new relation to plants. And what would be a better place to start than by taking a deep breath before beginning again?

30 December 2013–3 January 2014

3

SHARING UNIVERSAL BREATHING

[handwritten annotation: allergies cut him off from nature]

Since the age of three, I have been suffering from severe and debilitating "seasonal allergies." When one sunny day in May I experienced sudden facial swelling and was no longer able to breathe in the midst of a protracted asthma attack, my mother rushed me to a hospital, where the doctors diagnosed an acute allergic reaction. Every subsequent spring I have relived, on a smaller scale, the initial shock of those days that are etched among my earliest memories. Just when the revival of spring happens and the plant world begins to bristle with its radiant colors, I find myself cut off from the outside, barely able to see, smell, taste, or breathe. A watery barrier arises between my sense organs and their corresponding objects, disrupting the transparent workings of intentionality. Our physical and spiritual alienation from plants is, in the end, one and the same as our alienation from the rest of the organic and inorganic universe.

The rejuvenation of vegetation coincides with my autistic confinement within myself, complicated by an adverse reaction to the antihistaminic medicine, which is supposed to alleviate this condition. But I was not alone in my predicament, as most of my classmates developed seasonal allergies as well. There was no secret as to the reason behind the epidemic of pollen allergies: a large, air-polluting factory was situated just a few hundred meters from my neighborhood. In its infinite wisdom, the planning committee of the City of Moscow decided to construct this particular factory very close to the woods, so as to counterbalance the adverse effect of pollution with the air purified by plants. As a result, depending on the direction of the wind, we sensed either the smell of toxic fumes that emanated from the industrial monstrosity or fresh air that drifted from the woods.

Tree pollen and grass allergies, triggered by factory and traffic pollution, are symptomatic of everything that goes wrong with the human relation to plants. Often described as "the lungs of the planet," the woods that cover the earth offer us the gift of breathable air by releasing oxygen in the process of photosynthesis. But their capacity to renew the air polluted by human industry has long reached its limit. If we lack the breathing space necessary for a healthy life (or, indeed, for any kind of life), it is because we have both filled the air with chemicals and undercut the plants' own tendency toward regeneration: of their growth and of the atmosphere, which they enrich. Nevertheless, the warning that rapid deforestation combined with the massive burning of fossil fuels, which are largely the remnants of past plants, is an explosive recipe for an irreversible ecological disaster falls on deaf ears.

In harming the environment and, in general, the vegetal world, we are harming ourselves, as though blocking the oxygen supply that keeps us alive. However, when the body attempts to defend itself from undeniably dangerous toxins, it overreacts, shuts itself off from the outside, and rejects the nourishing and invigorating elements of the surrounding world. It thus misrecognizes benign plant pollen—rather than industrial waste that pollutes our air, water, and all the other elements—as the source of a threat. At the same time, allergies show that our extreme separation from exteriority, including that of plants, is untenable. After all, with every breath we take, we expose our lungs to the outside world, regardless of all the barriers we have erected between the environment and ourselves. Becoming completely isolated from the ecosystems we are a part of amounts to signing our own death sentence, given that without plants we cannot continue to breathe, live, or be.

In light of the allergies that have been inflicted on my body, I was determined not to let my thinking be affected by what I saw as the wide-spread conceptual allergy to plant life in Western philosophy. That is undeniably one of the motivations for my books on vegetal philosophy thus far. The more I worked on these writings, the more lucidly I saw the connection between the philosophical dogma that plants were the least advanced of living beings, a cultural rejection of any sort of ethical responsibility toward plants, and their economic transformation into raw materials for production or combustible biofuels. I realized that my physical allergies were caused, more or less directly, by the prevalent intellectual allergy to the complexities of vegetal life.

What, then, could be the way to recover our physical and spiritual breath, together with that of the plants themselves? On the face of it, everything in vegetal breathing is opposed to ours. Plants reprocess carbon dioxide into oxygen, while we breathe in oxygen and breathe out carbon dioxide; they breathe at the surface, through the leaves that are exposed to the outside, while we inhale by filling the interiority of the lungs with air. And yet, not all human breathing is accomplished by the lungs, seeing that our whole bodies breathe through the pores in our skin. We tend to pass over this more vegetal mode of respiration to which we are privy and concentrate, instead, on the hidden parts of our respiratory tract.

The European construction of spirit is grounded on this neglect of the breathing surface of the body and of plant respiration. As Emmanuel Levinas once said, spirit is "the longest breath there is,"[1] which means that what drives it is the desire to dominate the element of air, to incorporate as much as possible in the sublime and disincarnate lung of subjectivity, and to interrupt the sharing of breath with the outside world by indefinitely delaying the instant of exhalation. Is this pathetic model applicable to spirit *tout court*? Or, is it specific to the Western mutation of spirit, presumably separated from and oblivious to the needs of the body, which lives with, from, and through plants?

During one of my visits to India in the summer of 2009, a yoga teacher from the town of Kottayam in the southwestern region of Kerala insisted on the importance of practicing very slow breathing. He compared the life spans and breathing paces of a tortoise and a dog and then went on to link human well-being and longevity to a cultivation of breath. Was this advice parallel to the spiritual suspension of breathing in the West? A far cry from the "longest breath there is," the meditative practice of the teacher from Kottayam did not intend to appropriate the biggest share of elemental exteriority. On the contrary, he emphasized the need to let the breath pass through us better, with more awareness and attention. Similar to plants, we must become the conduits for air, channeling it through ourselves, rather than relating to it as a resource. I took the yoga teacher's suggestion not as a panacea to everything that is wrong with "the West" but as a different starting point, the other beginning, as Heidegger would say.

The process of breathing, scientifically understood as an exchange of gases between an organism and the atmosphere, is impossible without a true sharing between the inside and the outside. Modern humanity has unlearned the rules of such sharing and has appropriated much more than it has given back, not to mention that what it *has* dispensed to the surrounding world is little more than pollution and other deadly by-products of its "energy production." It has taken its own breath away: gasping with wonder at the sight of its technological achievements and suffocating under the fumes that emanate from them. More unfortunately still, it is robbing and blocking the breath of nonhuman forms of life, sacrificed on the high altar of spirit.

Restoring a more equitable exchange with the environment is the goal of most ecological programs and proposals for sustainable development.

But sharing is not reducible to exchange. We, together with everything that lives, always receive much more from plants than we give back to them, because plants breathe *for* all animal and human beings without exception. Although this would be an improvement over the current situation, the sharing of breath is not a matter of instituting a more just natural economy. It is, rather, the issue of acknowledging the unpayable debt we owe to plants, of partaking of the generosity in which plants excel, and of drawing inspiration from this generosity with every breath we take.

If the breathing process is elevated to the level of a mindful practice, then nature is cultivated, both within us and outside us, instead of being enucleated and destroyed. Such cultivation is an alternative origin of culture that does not break its bond to the environment and does not aim to dominate the vegetal world. It recognizes that all terrestrial living beings breathe within the same atmosphere, to which they belong, in which they are rooted (Heraclitus, for instance, considers breathing to be a root, *riza*, that tethers us to our surroundings—cf. Sextus Empiricus, *Adversus Mathematicos* VII, 129), and which they have to share in order to subsist. What differs is how each one receives this gift and what each one can do in order to support this fragile world.

This brings me to the sense of sharing, to be carefully distinguished from "dividing." The difference between the two is that *sharing life and breath augments and enhances the sphere of the living, whereas dividing life into the so-called natural or human resources diminishes it.* Life is what happens between two or more, in contrast to the division that leaves no space in-between the living. Division is, therefore, inherently suffocating, because breathing can only happen in the interstices, when a minimum of "breathing space" is guaranteed. Our task is to develop a culture through which we would come to view the plants, ourselves, and the air between us as the participants in the ensemble of life, rather than a mesh of quantifiable objects or productive potentialities. That is to say: a culture not allergic to life but flourishing with it.

The cultivation of sharing, instead of division, is perhaps the sole practical and theoretical trend that can still make a positive difference in the disastrous environmental situation we find ourselves in. And where is sharing to commence if not in the rudimentary activity of our living bodies that necessarily breathe together with (and also through) others, notably with and through plants? When, for weeks, if not months, I suffer from pollen allergies, I am unable to practice my breathing, because every breath is a struggle. The same is true for our autistic, allergic culture that shuns life. Having undermined existence so much that everything meaningful pales in comparison with the goal of mere survival, it has deprived us not only of the space but also of the time for deliberate, mindful practice, whether breathing or any other. Within the mix of industrial pollution and spiritual suffocation, how to find the time and the space for breathing and, hence, living otherwise?

On the other hand, the plain fact that plants still regenerate every spring is highly significant. The hint for a solution resides there, in the contradiction between cultural suffocation and the natural respiration of plants. We can perhaps begin to change our breathing patterns in the green spaces where division and appropriation do not play a significant role, unlike sharing. After my seasonal allergies are over, at the dawn of another season, I can return to the woods and finally open my lungs to the freshness of vegetal respiration. In being with plants, everything slows down to the pace of their growth, and it is in this slowdown, echoing the suggestion of the teacher from Kottayam, that another cultivation of life is realized.

18–24 January 2014

4

THE GENERATIVE POTENTIAL
OF THE ELEMENTS

A return to the classical notion of the elements (in Greek, *stoicheia*) is not only a strategy for confronting the prevalent modern attitude, which Max Weber has called "the disenchantment of the world." If this were the case, the thinking of the elemental would have been but a reaction to the scientific rationality that understands the world on the basis of mechanical causes and effects. Instead, revisiting the elements may help us uncover a different kind of logic compatible with life.

Across various traditions, the classical elements usually include air, water, earth, and fire. Jain thought, additionally, understood plants to be the fifth element comprising the universe, alongside the four I have just mentioned. Plants are, at the same time, ensouled beings, *jivatthikayal*, and the sources of the world. In the Jain *Ācārāṅga Sūtra*, the Venerable Ascetic Mahāvira came to the realization that "the earth-bodies and water-bodies and fire-bodies and wind-bodies, the lichens, seeds, and sprouts" are "imbued with life, and avoided injuring them" (1:8:1:11–12). It is, precisely, the generative capacity of plants—their seeds and the sprouts, to which they give rise—that elevates them to the status of the elements. Better yet: the earth-bodies, water-bodies, fire-bodies, and wind-bodies are added to the elemental growth of plants that reveal how the entire world lives, is full of countless *jivas* or souls, and flourishes.

Among the pre-Socratics, Empedocles is especially sensitive to the vegetal nature of the elements, to which he refers as "the primordial roots" (*rizomata prōton*) of all things (Fr. 6, Aetius I, 3, 20). The world that grows from these roots is certainly not static; constantly shifting and changing, the elements are combined into the one from the many by the power of love and fall apart again under the influence of strife (Fr. 17, Simplicius *Physics* 158). Nothing can be generated without love, which resides in the midst of (*en toisin*) the four roots, which is responsible for growth, and which brings with it the measure, harmony, and balance conducive for being. Strife, on the other hand, introduces—from the outside, separately (*dicha tōn*)—imbalance and division into the relations among the elements, forcing provisional unities to fall apart. Vegetal growth, the model for everything that springs into existence thanks to a harmonious joining of the elements, is the work of love, while decay is the product of strife.

The four roots, together with the two powers of love and strife, are equally the sources of living beings and of their capacity for perception: "For with earth do we see earth, with water water, with air bright air, with fire consuming fire; with love do we see love, strife with dread strife" (Fr. 109, Aristotle, *Metaphysics* B4, 1000b6). The elements generate both physical and spiritual life, the one encountering the other by means of the particular root they share. The compositions and the proportions of the four roots in the soul reflect and reiterate their composition and proportion in the outside world. The soul opens to exteriority by means of the elements of that very exteriority; from the first moment it is inhabited by the earth, which inhabits and welcomes within itself the atmosphere it is surrounded by. But, if Empedocles embraces a certain vegetal model of generativity in the case of this world, then he must subscribe to the same model in the case of the soul. It follows, then, that the basis of psychic life, like that of the growing physical world, is vegetal. Restated, this is one of the conclusions of *Plant-Thinking*, where I direct the readers' attention to the emanation of the seemingly abstract and sublimated thought processes from the vegetal "substratum" of the human psyche.

To a certain extent, plants are even more elemental than the elements themselves, in that they "exhale" breathable air, indispensable to the existence of all other beings. By adding oxygen to the air, they make air what it finally is: an element propitious to breath and to life. It was, of course, Anaximenes who postulated that air was the main element that, by condensation and rarefication, was transformed into all the others (Simplicius, *Physics* 24, 26). Hence, plants make possible the element that makes all the rest possible; though they are generated beings, plants also generate the entire world.

The relation between the air within and the air outside us, between psychical and cosmic breath (*kosmon pneuma*), is crucial for Anaximenes. The body of a living being and the body of the world are gathered into their respective ensembles thanks to the air/soul that encloses them and that they enclose (Aetius I, 3, 4). Although it is a matter of debate which words in this fragment are actually those of Anaximenes, what this implies is that the body and the world each receive their coherence, their order, and their vitality from the breath appropriate to them. The world is analogous to a living being nourished by air, which, in the form of a soul, guides the bodies of plants, animals, humans, and the entire universe. A body that is ready to *receive* air is a living one: indeed, this receptivity, which is nothing substantive or identifiable, includes it in the order of life itself. How intensely living, then, is the body of a plant that *gives* breathable air!

Under the influence of Aristotle, the elements of the pre-Socratic thinkers have been (formally) conceptualized as the first organizing principles of everything in the world. Plutarch reports, for instance, that water was such a principle or beginning (*archen apantōn*) for Homer and Thales (*de Iside et Osiride* 34, 364D). Aristotle notes in *Metaphysics* that

Thales deduced the primacy of water for life from the fact that all things that came into being required moisture to germinate, since "the seeds of all things [to pantōn ta spermata] have a moist nature" (Metaphysics A3, 983b6). In other words, water awakens seeds into germination from the dry state in which they are lifelessly preserved. Plant seeds united with water are, in effect, exemplary here with reference to the elemental origination of all other living beings or "the seeds of all things."

This observation leads Thales to far-reaching insights on the nature of the soul and of the world. If water animates seeds, calling them forth into existence, then it must harbor the principle of animation, or the soul. Hence, Aristotle reports, some "thinkers actually declared it [the soul] to be water . . . they seem to have been persuaded by the seed of all things being moist" (De Anima A2, 405b1). The soul is not a separate invisible "substance," distinct from the stuff of the outside world; it is made of the very element that comprises everything by condensation and rarefaction. The soul is the most rarefied, vaporized water there is. At the same time, life-giving water is the substratum for the earth that "stays in place by floating like a log" (Aristotle, De Caelo B13, 294a28). The earth, which is condensed water, does not drown like a stone tending to its proper region beneath, but stays afloat like a piece of wood. So the entire earth is vegetal in relation to the water that subtends it.

What of the fire of Heraclitus, then? Can this uncreated, "ever-living" element, whose measured (metra) kindling and going out makes the entire cosmos or world-order (Fr. 30, Clement, Stromata v, 104, i), really burn without something that fuels it, namely what we now call "matter" and what, more simply, referred to wood (hylē) in Aristotle's time and prior to that? Isn't a fire without the materials for combustion impossible? To be sure, those materials are not limited to parts of plants that humans have burned along the course of their history to keep warm, cook their food, or, later on, power their technology. The Heraclitean fire burns both outside and within us—as us, too—at the same time bestowing life and consuming or depleting it. That is the enigmatic sense of the final words in Fragment 62, "living their death and dying their life" (Hippolytus, *Refutatio omnium haeresium* IX, 10, 6): as our inner fire continues to burn, it grants life, which is burning *out*. In the spirit of Heraclitus, one might say that we are the wood in the fire of life, which, unlike that of the universe, is finite.

Unfortunately, millennia after Heraclitus, the element of fire has become more destructive than creative. As I write in *Pyropolitics*: "When a physicist conceptualizes matter as accumulated and temporarily held-back energy; when we view our diets through the lens of caloric intake, and fitness—in terms of calories to be burnt; when the quest for alternative sources of energy leads governments seriously to consider the prospect of burning *anything* whatsoever, to accelerate deforestation, and to spread plant monocultures (such as sugarcane in Brazil) for the sole purpose of transforming them into biofuels: when all this takes place, then fire comes to dominate our sense of reality. Perhaps, life itself is an internal conflagration, a great fire in which all living beings are so many sparks, igniting other similar sparks in reproducing themselves. We would not swerve far from the ancient Greek take on the life-giving power of heat in making this assertion. But, while for the Greeks, the creative potential of fire had to do with its measured, controlled, periodic lighting up and extinguishing, for us all sense of measure has been lost as the blaze rages uncontrollably. As the worldwide fire grows, so does the destruction."[1]

Moisture and heat, water and fire, which are incidentally treated as "the elements of opposition" in Hegel's *Philosophy of Nature*,[2] need to be present in the right proportion to be conducive to life and, in the first place, to the life of plants. Too much or too little water and light (as well as heat) is detrimental to vegetal germination and growth. Therefore, the elements carry a generative potential within strict limits that are congruent with existence. Outside these limits, when the elements are either too abundant or too scarce, a living being is no longer able to engage with them and, consequently, cannot continue to exist. The measure of the elements propitious to life—our own and that of plants—is exceptionally fragile. While plants contribute to its maintenance, humans have been undermining it by altering elemental limits on the planetary scale, for instance, through industrial activity leading to global climate change. These changes have resulted in some parts of the earth becoming too dry (e.g., the spread of the Sahara desert), while others, such as England, are wetter than they have ever been. The extremes of severe drought and flooding make those places no longer welcoming toward the life of plants, as well as animals or humans. Elemental dearth and overabundance are the two extremes between which existence is finally possible.

Despite the theoretical quarrel among pre-Socratic philosophers, regarding the exact element that is more primary or more important, it is the combination of all four elements, in the right proportion, that holds the potential for germination, growth, and other manifestations of life. *That* plants exist, thriving in a certain place, is evidence enough for the coming together of the elements within the limits appropriate for being. A plant is such a gathering place for the life-giving elements, neither too overpowering nor too weak. Its thriving betokens the gentleness of elemental nourishment: sufficient sunlight, moisture, minerals in the soil, and air. We cannot avoid mentioning that the latter two elements— the earth and air—are themselves enriched by the activity of plants that, by decaying, replenish the fertility of the soil and, by photosynthesis, introduce more oxygen into the atmosphere. In the right proportion, the elements extend hospitality to vegetal existence, which, in turn, *nourishes the elements* and convokes other kinds of vitality. No wonder that certain traditions, like Jain philosophy, have included vegetation as the fifth element of the world. After all, plants contribute to the excellence of the "classical" elements by augmenting their generative potential and making them more propitious to life's flourishing.

27 February–2 March 2014

5

LIVING AT THE RHYTHM
OF THE SEASONS

Twentieth-century Russian philosopher Vladimir Bibikhin often used the word *neumestnyi* to describe the nature of a human being. In English, *neumestnyi* means "unsuitable," "strange," "uncanny," though, more literally, it signifies "the one who is out of place (*mesto*)" or "the one who has no place," "is not in a place." In assigning their proper places to all living and nonliving beings, Bibikhin contends, the humans lack a determinate, proper, or strictly circumscribed place of their own. They paradoxically stand outside the totality they laboriously put together *from within*.

The lack of a human place in the living whole puts us at the greatest remove from the world of plants that are rooted in the earth. I sense the history of my serial uprootings as the intensification of this tragic condition and I turn to vegetal life, both in thought and in my practice of living, in order to remedy it to the extent possible. But the etymological intuition in Russian goes further than that: without their place, human beings are no longer suited to their environment, to the world that they physically inhabit and that they, in turn, make less and less suitable for life. The metaphysical search for another place—indeed, for the other *of* place—beginning with the Platonic *topos ouranios*, or the heavenly sphere which would override in its significance everything in this world, is the side effect of that original displacement.

Not only do humans tend to live out of place but they also seem to exist out of time, noncontemporaneously with the natural world and with themselves. Symptomatically, Nietzsche's *Unzeitgemässe Betrachtungen* can be rendered in English as *Untimely Meditations* (Walter Kaufmann's and Hollingdale's alternative)[1] or *Thoughts out of Season* (the option preferred by A. M. Ludovoci and Adrian Collins). Furthermore, for the age of nihilism, Nietzsche predicted the kind of culture that would be a "fruit out of season": "every season has its own particular charm and excellences and excludes those of the other seasons. That which grew up out of religion and in proximity to it cannot grow again if this is destroyed; at the most, stray, late-sprouting shoots can put in a deceptive appearance."[2] Nihilistic culture suppresses the origination of culture in the work of cultivation, for the most part exerted on vegetal nature, the work that demands patience, the capacity to wait for the crops or for the time the plants themselves require to mature, and attunement to the changes of seasons. Its "fruit out of season" is wholly inappropriate, improper to the times (or timing) and places of growth.

The word *season* harks back to the human engagement with plants, seeing that it is derived from the Latin verb *serere*, "to sow." Various seasons are the best times to sow, to care for, and to reap, for instance cereals or flowers. While seasonal changes (expressed in the fluctuations of temperature, humidity, and the amount of daylight) closely follow the rotation of the earth and its position vis-à-vis the sun, plants track these changes and take them as cues for blossoming or coming to fruition, remaining dormant as seeds in the cold dark of the soil or sprouting to the warmth of spring. In other words, the seasons connote an alteration and an alternation: the becoming-other of the summer in the fall and the winter and the cyclical nature of change, when winter, too, is followed by the spring and the summer. All this depends on the place where living beings are in relation to the equator: equatorial climates demonstrate little difference among the four seasons and, as a result, most plants growing there remain evergreen all year round.

To live out of season is to ignore the alterations and alternations of planetary time and to exist out of tune with the milestones of vegetal time: germination, growth, blossoming, and fruition. What is worse is that human technology has managed to regulate seasonal and diurnal changes in order to stimulate the growth of plants for consumption. A hothouse is a rudimentary instrument that interferes in the environmental conditions of plants to encourage their "production" year round. Often, commercial growers subject plants to supplemental artificial light in order to speed up their growth, extending the photoperiod to sixteen, twenty, and at times twenty-four hours a day. Like humans, plants can be violated by uninterrupted light that forces them to grow without nocturnal, as well as seasonal, breaks. The leaves of corn, for example, show signs of damage after the plants are denied the cyclical time of alteration and alternation—of light, heat, and so forth.

To put it differently, when seasonal (or daily) cycles are disrupted, what results is a certain arrhythmia, a disruption at the level of being itself. Western culture in late modernity takes pride in this effect, equating it to a triumph over nature and to the freedom of overcoming the limits (perceived as constraints) of the outside world. But it is these very limits, understood by the ancients as the inherent measures of being, which need to come together in precise proportions in order for the being to be, that potentiate life and existence. The unraveling of the measures of vegetal time, dependent upon the reception of the first and the last rays of the sun in the photosensitive cells of plants, is tantamount to the unraveling of their (and our) being. Moreover, this fatal process is now happening on a planetary scale due to global climatic changes induced by centuries of human industry and pollution. The relentless rains that have been falling on England, Portugal, and Spain this year have upset the deepest sense of seasonal existence. The normal February sowing period in Portugal was not propitious to the sprouting of seeds because the earth was too saturated with moisture, which caused the seeds to rot. By robbing plants of their time, and especially of their future, we deny a future to human and all other living beings.

At the origin of the seasons, then, planetary time is measured by the stages of plant life and of the human dealings with plants. There is an appropriate time for entrusting the seed to the earth—the season of sowing, which came to be a synecdoche for all the other seasons. Another season arrives when the young sprouts reach toward the airy expanse of the sky, partly leaving the darkness of the soil, or when plants blossom. There is also a season for ripening, absorbing the light and the heat of the sun (fire), bearing fruit, and harvesting. And, finally, there is a season of rest, of being covered by snow or returning to the waters that fall from the sky. The reason why I am casting seasonal alteration and alternation in these terms is that each season demonstrates a specific relation to the elements, one of them becoming definitive and playing a more important formative role for plants and for vegetal time. The seasons not only convey the journey of our planet around the sun but also an elemental cycle, a procession of the elements over the year. Living at the rhythm of the seasons means respecting the time of plants and, along with them, successively opening oneself to various elements.

In the biblical Ecclesiastes, this rhythm has been already corrupted, as the heavenly element comes to prevail over the rest, the time and integrity of plants are disrespected, and murder, destruction, and violence enter the fray. "To every thing there is a season, and a time to every purpose under the heaven [*shamayim*]," it begins (Ecclesiastes 3:1). Although the idea of the appropriate time is preserved in this programmatic statement about the seasons, every purpose and activity is gathered under the sky (*shamayim*) that presides over the other elements and, thus, skews the balanced rhythm of seasonal alteration and alternation. Everything

is polarized here; there are only two options, the positive and the negative: "A time to be born and a time to die; a time to plant and a time to
uproot [la'akor] what has been planted; a time to kill and a time to heal;
a time to break down and a time to build" (Ecclesiastes 3:2–3). Should we
be surprised with the melancholy, nihilistic conclusion of Ecclesiastes,
"That which has been is that which shall be, and that which has been
done is that which shall be done; and there is nothing new under the
sun" (Ecclesiastes 1:9)? After all, highly polarized binary actions have
eliminated natural diversity, simplified the seasons, and subjected everything to the tyranny of that which presides over this monotonous rou-
tine, be it heaven or the sun. Vegetal seasons are reduced to planting or
sowing and the excoriation of plants, without leaving any time for care
or cultivation. In fact, much of the beginning of Ecclesiastes is devoted
to denying any difference, or creative potential, to the elements: the sun
and the wind—fire and air—travel around the earth in seemingly point-
less circles, rivers run toward but do not saturate the sea, generations
pass and the earth remains eternally the same (Ecclesiastes 1:4–7).

It is only from this incipient metaphysical point of view that the
rhythm of the seasons assumes the shape of a meaningless repetition.
Difference—alteration and alternation—is assimilated to sameness, to
which even the elements are reduced. Growth, generation, becoming,
and blossoming are seen as inessential in comparison to the firmness of
the earth and, ultimately, of being. But the rhythm of the seasons is far
from monotonous. Each time anew the earth regenerates; every plant
that sprouts is unique; each generation—unrepeatable.

Nor is it a matter of indifference when human beings are born and
when they die. I am not referring so much to astrological charts and calculations as to the sense that the moment of death always arrives too
early, is premature. Elsewhere, I have discussed in more or less detail
Heidegger's assertion that human life does not come to fruition in death
and therefore does not accomplish its purpose (*telos*) in the manner of a

plant that bears fruit in an appropriate season. Death always comes out of season for us, which is why we tend to live out of season, too, leading a life that is inflected by the rejection of mortality and finitude. Particularly telling is the comparison of the human end with the time of the harvest: the seasonality of vegetal processes contrasts very sharply with the unseasonable course and end of human existence.

A life out of season is one of empty, unfulfilled desire, that is, a life of nonsatisfaction. Lucretius puts this well in *On the Nature of Things*, where he writes that "the seasons of the year . . . come around with their fruits and manifold delights and yet never satisfy our appetite for the fruits of life (*vitai fructibus*)" (III, 1000–10). Conversely, living with the rhythm of the seasons is delighting in what each of them has to offer, without wishing for more: the warmth of spring or the crisp lucidity of autumn, nourishing moisture or abundant sunlight, the activation of growth or rest. It is from plants that we can learn such fulfillment, for instance when we visit the same place in the woods all year round, in different seasons, and observe how the trees growing there adapt to the changes of weather and the elements prevailing at the time. I do so on my daily walks in the pine forest, situated directly behind my house, and I delight in the intense smell of pine resin in the heat of summer, the fragile wildflowers that blossom between the trees in the spring, or the wet freshness of the woods in autumn and winter.

Overall, living at the rhythm of seasons requires figuring out an appropriate way of being and behaving, given the conditions we find outside ourselves. As such, there is no fixed recipe, but only a repeated practice, aiming at what is appropriate. Seasonal living also entails, without a doubt, seasonal eating: enjoying the fruits gifted to us by the vegetal world at their own time of fruition. During the summertime in Portugal, one anticipates seasonal fruit, starting with strawberries and ending with persimmons, before a long period of citruses, such as oranges, limes, mandarins, etc., that are widely available in the winter. On human whim, it is possible to grow strawberries in hothouses virtually all year or to import them from another continent (there are no seasons for globalization!), but eating these berries over the course of the year does not amount to living at the rhythm of the seasons. In fact, it is a sign of disrespect for the inherent, enabling limitations of the time of plants and of the environment. Seasonal living means, then, adjusting our needs and desires to what the vegetal world furnishes at any given time to our vision and touch, olfactory sense and taste buds. By bringing us closer to planetary and elemental time, it allows us to approximate, if only slightly, the time of plants.

22–26 February 2014

6

A RECOVERY OF THE AMAZING
DIVERSITY OF NATURAL PRESENCE

My hypothesis, which is both a theoretical inkling and a personal hunch, is that we begin to feel a sense of wonder before the diversity of nature especially in those periods when our lives in built environments become virtually unlivable. In Hegel's *Phenomenology*, Spirit (*Geist*) is at home with itself in the absolute, on the basis of which the entire world will be reconstructed despite the disquietude of self-negating thought.[1] Comfortable and secure in this transformed reality, dialectics can afford to dismiss the rich variations of sensory experience (and of natural givenness, which is not deliberately *self*-given) as belonging to the poorest and most abstract stage of Spirit's journey. In order for our prejudicial relation to nature to change, the illusion of comfort must dissipate. Rather than treat the worlds of the elements, vegetation, and animal life as threats, we can turn to them for the lost sense of well-being, rediscovering its diversity, irreducible to a single metaphysical principle, including that of Life.

As a doctoral student of philosophy in New York City, I rented a small room in the basement of a building, located in Manhattan's East Village. More of a narrow corridor than a room—with exposed pipes, a small bed, and an improvised tiny shower cabin—this "studio" culminated in a big floor-to-ceiling window that opened onto a spacious backyard of the kind that one would not suspect existed in the midst of the city's concrete jungle. It was a musty room, incredibly suffocating both due to its low ceiling and the smell of cigarette and cigar smoke with which its walls had been imbued over the years. "Focus on the positive," the landlady said, charging me rental prices I could hardly afford. "Your new home is all about what is outside, not the inside." She, herself, probably did not know to what extent she was right. At my new address (East 10th Street, between Avenue A and First Avenue) I had to learn what living and think-ing outside meant. And, to refer to a line by Portuguese poet Fernando Pessoa, I had to take "lessons in unlearning" when it came to the com-forts of everyday existence I had hitherto taken for granted.

Even if the world of plants held a special significance for me before that period of my life, at this time it became completely indispensable. Given the minimalism of the crammed room I inhabited, the single ground-level window that opened onto a lush backyard became my preferred escape hatch. For me it was more than a simple window, as it led to a quiet place in nature (so near yet so far), where I could breathe, live, and revel in the rich splendor of colors and scents that were absent from the dark and stuffy room. I would spend hours in the yard, seated at a half-rusted iron table, and alternate between reading and imbibing the sights, smells, and rustling sounds of vegetation there: the tall, untrimmed grass; the dense shrubs; an ailanthus, commonly called the Tree of Heaven, that must have been over a century old . . . The size of the bustling yard contrasted so starkly with the dingy environment of the room that there was no doubt which one was more welcoming toward living and thinking. But exactly how was it possible to recover a different experience with and in nature?

One thing is incontrovertible: such a recovery did not (and does not) result from a reconstruction of experience "from scratch," as though we were to start again from a blank slate. Both Georges Perec in "Kinds of Space" and Gaston Bachelard in *The Poetics of Space* endeavor gradu-ally to rebuild lived space,[2] or a sense of place, from the most basic constituents either by emptying the dwelling altogether (into a kind of nothingness in Perec) or by starting from its smallest recesses (as in Bachelard's "house"). Broadly speaking, this tactic requires the initial reduction of the world with a view to its subsequent recovery on a page or on the plane of thinking. But the page is, itself, the afterlife of a tree, which means that there is nothing blank or ideal about its white space. And the same is true for thinking! Phenomenological or quasi-phenomenological reduction, which is extremely anti-environmental, cannot do away with what animates life. Being in or with nature excludes a return to the much-fantasized "clean slate" of the philosophers. All we can do is pick up a lost (but also ever-present, in the deep background of our lives) thread of existence.

The efforts at reconstituting diversity and difference, especially those of vegetal and other kinds of nature, on the basis of an initial emptying out or evacuation of the world are bound to fail. Our recovery of the experience of natural diversity has no luxury of a gradual, meticulous, step-by-step reconstruction; if it is to happen, it must come to pass all at once. Crossing the threshold between the walled-in environments we normally inhabit and the open places where plants grow, we emerge from the one to the other instantaneously. Similarly, the diversity of vegetal life strikes us forthwith, without warning. After passing the threshold, which, in my case, assumed the shape of a ground-level window, we can no longer resort to artificially constructed limits, boundaries, or walls, whether conceptual or physical, for delimiting life. On the hither side, a single tree is already an ensemble of multiple growths: divergent and interweaving trunks; the moss or the ivy that covers the branches and a squirrel that climbs them; the birds that weave their nests there; the microorganisms that dwell on and in vicinity of the roots . . . A community of growths, the tree is not only vegetal. It is a meeting place of the elements, vegetal forms, species, and biological kingdoms. It gives itself to sight and to thought together with everything that dwells on, under, and above it—and with the place where it dwells. A tree, too, can be a window unto the unfamiliar realm of nature that is oblivious to classifications, boundaries, or principles.

What the impossibility of reduction really means is that there are no "pure" colors, scents, or sounds in nature, distilled from the living beings who display or produce them. In the garden I see the green *of* this grass, smell the scent *of* these flowering shrubs called *Cloisya*, and hear the wind touching the leaves and branches *of* this olive tree. Just as there are no stimuli separate from the sensuous surfaces of the world (let alone of the vegetal world), so there is no "pure" vision, hearing, touch, scent, and so forth, awaiting me upon my return to something like natural presence, the vibrant *parousia* of nature. Instead, my sensorium is invited to plunge into, to submerge itself in the world of the elements and of plants, without the excitement of overstimulation, but rather with a serenity that envelops the body.

From the standpoint of rational metaphysics, the paradox of the recovery I have described as faithfully as I could in the foregoing is that it finds simplicity beyond the monotony of sameness and outside any postulation of "first principles." The simplicity of difference and diversity has to do with my lingering in a singular place, not wishing to leave it in a rush to go somewhere else, and also with my experience of time in that place, the time that does not hasten to transport me to the past or to the future beyond the current horizon. The closest analogy to this experience, perhaps, is the feeling of one's own body—feeling oneself feel. The experiencing and experienced body is not given as an undifferentiated object; it is felt as a whole, in the simplicity of existing, across the diversity of its surfaces and depths, powers and receptivities, foldings and unfoldings. When Marx provocatively calls nature "man's inorganic body," in an expression from which Deleuze and Guattari have derived their "body without organs," we should not hear in his words an appeal for a better domination of this strange corporeality by the collective mind of humanity. If we wanted to be charitable to Marx, we could say that an experience of nature outside the mediations of exchange-value is similar to the experience of our own bodies rescued from the force of commodification: both are the simplicities of infinite differences that can be occasionally, in the most felicitous moments, sensed all at once. Furthermore, the simple diversity of the body is felt most vividly thanks to the other, with whom it is discovered or recovered, for instance, in love. Outside, in a backyard or in a garden, a richer sense of one's body, as much as of one's thinking, is also achieved in contact with the other, with the elements and the vegetal world.

There is not—there cannot be—a clear method, a tried-and-tested way toward such an experience of gathering without abolishing difference. I cannot give readers a recipe: everyone must find her or his window into the backyard for herself or himself. Indeed, if I could speculate about some general patterns of intuiting this openness, I would say that its outlines vary depending on whether it is "her or his window." Even the meaning of the possessive pronouns, *hers* and *his,* will not be the same for everyone, for instance, if, in the simplicity of infinite difference, she *or* he perceives *or* they perceive that this window opens in her *or* in him *or* in them and that it, this window, *is* her *or* him *or* them. The diversity of natural presence supports equally diverse approaches to and points of entry into the world of plants. The question is how to frame a window within a window, namely the windows that we are and the windows that each plant opens onto a much wider field of life and experience.

To a metaphysical way of thinking such a proliferation of differences is anathema. In place of multiple approaches, metaphysics postulates a single correct method that supposedly culminates in an objective truth so long as we follow it to the letter. In place of the diversity of plant life, it posits a unified principle orchestrating everything from behind the scenes. For Goethe, this principle is a part of plants, from which all others can be derived or to which they can all be reduced. As he specifies in

The Metamorphosis of Plants, the basic building block of plants is the leaf, rarefied into fragrant petals or condensed into the seed. (Doesn't such rarefaction and condensation remind us of Thales with his thinking of the metamorphoses that water undergoes in the generation of everything?) In a poem that precedes his careful botanical study, Goethe starts with a sense of bewilderment at the variety of vegetal forms: "The rich profusion thee confounds, my love, / Of flowers spread athwart the garden. Aye, / Name upon name assails thy ears, and each / More barbarous-sounding than the one before—Like unto each the form, yet none alike; / And so the choir hints a secret law."[3] Very quickly, however, a formulation of "the secret law" cuts through the confusion. When the whole is "reflected in each separate part," it is time to "Turn now thine eyes again, love, to the teeming / Profusion. See its bafflement dispelled. / Each plant thee heralds now the iron laws."[4] A "teeming profusion"—that is how metaphysics reacts to the manifold of vegetal differences. And the role of "iron laws" is to put an end to the wonder (the "bafflement") that we experience in the face of nature's diversity.

Beautiful as his poetry might be, Goethe makes for a poor field guide to plant world. Much more respectful of vegetal life in all its irreducible variety is the student of Aristotle, Theophrastus, who begins his *Peri photon historias* (*Enquiry Into Plants*) with a list of *aporiae* (translated by Arthur Hort as "difficulties") making it impossible to establish with any degree of certainty what essentially belongs to a plant and what doesn't (I.i.1–2). Compared to an animal, the number of plant parts is "indeterminate [*aoristos*] and constantly changing" (I.i.2), "for a plant has the power of growth in all its parts, inasmuch as it has life in all its parts [*pantaché zōon*]" (I.i.4). Theophrastus gives us to think a genuine metamorphosis that does not reach closure and completion in any plant part, least of all the leaf. Vegetal life retains its indeterminacy, vibrancy, and diversity.

"In fact," Theophrastus continues, "your plant is a thing various [*poikilon:* multicolored, as in a tapestry] and manifold, and so it is difficult to describe in general terms: in proof whereof we have the fact that we cannot here seize on any universal character which is common to all" (I.i.10). Variety and manifoldness are so engrained in every single plant and in its parts that a general description, gathering these features into the unity of a concept, verges on being unfaithful to that which is described. The best we can do, at the level of a more abstract thought, is repeat after Theophrastus: the plant is an indeterminate and undeterminable sum of its differences (*diaphora*). This statement is accurate, but incredibly hollow. To fill it out, it is necessary to come out of the four walls that surround us and to expose our senses and thoughts to the differences that the plants, every single plant and every part of a plant, are.

27 February–14 March 2014

7

CULTIVATING OUR SENSORY PERCEPTIONS

Unjustly considered insensitive by philosophers and public opinion alike, plants by far exceed animals and human beings in their attentiveness to what is going on around them with regard to the levels of light, heat, moisture, movement, vibration . . . They are constantly in touch with the elements. Scientists have been for a long time familiar with the sensitivity of vegetal life, though in most cases they have proceeded to utilize it for purposes that are extraneous to the plants themselves.

A tree can detect an impending earthquake days in advance because of its roots' attunement to the minute vibrations of the subsoil. Human-made implements are capable of translating such vegetal detection into warning signals that are meaningful to us. Is this how we should pay attention to the attention of the plant? Must we convert its sensitivity to the outside into a set of data in order to help us live or survive better (here: by evacuating a potential earthquake area ahead of time)? If the horizon of possibilities were reduced to this option, wouldn't we treat plants as highly potent instruments, measuring devices, or *antennae*—a word consistently used in scientific literature—and thus lose sight of everything that is living in them?

Humans tend to pay attention to nature when the elements do not cooperate with us, upsetting our plans. As soon as something goes wrong, bad weather strikes, the earth literally gives way underneath our feet, or the sky falls on us, we direct our gazes up or down, as the case may be, and, traumatized, recall our inclusion in the fold of the earth, our dwelling place.

The Great North American Ice Storm of 1998, which struck Ottawa in my first year there, made residents realize to what extent their lives, the infrastructure of their cities, and especially the "power grid" are still integrated with the elemental forces and the world of plants. The storm system left between six and ten centimeters of ice accumulated on branches and power lines, toppling trees and causing weeks of disruptions and electricity outages amid freezing temperatures. Those of us who went through that ordeal could not ignore either the elements or the limits they imposed on the human mastery of nature. All this would seem to suggest that Heidegger was right, and we are shaken back to life only upon looking death in the face. But another kind of attention, one that is not oriented to death, is also conceivable and, moreover, necessary for our lives and for those of plants.[1]

Being under a direct sway of the elements, unprotected by the entire apparatus of civilization with which we shield ourselves from our environment, is perhaps more than we can bear. What is particularly difficult to come to terms with is the—real or perceived—indifference of the elements to individual existence. We are dwarfed (if not crushed, swallowed up, or dissolved) in their "sublime" milieu. To avoid the feeling of inferiority, we have separated ourselves from them by artificially constructing a world where we can feel at home. Now, the walls of this world are so high that we have ceased noticing the elements and considering how they constitute and support our existence. We can either wait for these walls to collapse in the aftermath of a catastrophe, brought about by the current culture of death, or we can scale them and build a different dwelling on the other side, one more open to the environment.

By analogy with how, during the hottest summer months, we seek protection from relentless sunshine in the cool shade of a tree, we can turn to plants for an intervention in our relation to the elements. It is through vegetal life, both within and outside us, that we can cultivate a nontraumatic mode of paying attention to the elemental.

I have always preferred sunshine sifted through the lush crown of a tree to the intense light that strikes the eye on the seashore, where brightness is magnified by the white of sand and the reflection of the rays on the surface of the sea. In the woods the glow of the sun is softer, subtler, more playful, with intricate mosaics of light and shadow. Plants maximize their exposure to sunlight, which they welcome with their leaves. They attend to the sun with every part of their bodies, proliferating in the direction of the celestial blaze. Humans can also bathe in light—and not only of the physical variety!—but we do not extend unconditional vegetal hospitality to this element. Nor do we attend to water, the earth, or the sky as plants do. Our attention span is shorter, receptive capacity weaker, attachment to whatever or whoever we attend to less faithful.

Seeking refuge, we flee from the elements, while plants expose themselves to the air, the earth, and moisture. It is this exposure that marked the human condition in the Garden of Eden where life was more distinctly vegetal. After the Fall, Adam and Eve used the leaves of trees, the most exposed parts of plants, to hide their own nakedness or exposure. With that gesture, they interposed the first screen between themselves and the world, between the two of them, and between each and her or his self. They lost the paradisiacal ability to attend to oneself and to the others in absolute openness. That is to say, they interrupted the flows of energy between psychic and physical interiority, on the one hand, and exteriority, on the other. As a result, their attentive comportment suffered irreparable changes and their distance from the elements grew.

When I linger with plants, in thoughtful and physical proximity, I try to pay attention to their singular mode of attention. I notice, first, that plants do not attend to an *object* or a *group of objects*. Their attention is inseparable from their life and growth. From a magnificent sequoia to a blade of grass, a plant attends to the physical elements, precisely, because the elements are not objects and cannot be objectified. Only then, in such nonobjectification, are the elements and life itself respected in their proper being. Therefore, human attention convoked and directed toward life must strive, strange as this may sound, to be similarly nonobjectifying.

Second, I appreciate the vegetal way of attending to the elements *by contiguity*. Plants are together with what they attend to, and their being is being-together with air, moisture, soil, warmth, and sunlight. In their attention to the elements, they become themselves. They draw their difference and uniqueness from the apparent elemental indifference. When I linger with plants, I find myself thus in a communion with everything they are and live with. And I am together with myself differently as well; I become myself, otherwise.

Third, it seems to me in the moments of silent meditation in the woods that plants pay attention to the elements *continuously*. Their alertness is not awakened solely when their lives are endangered (as ours is when we cross a busy road), which means that it is not summoned by death. Life processes, such as breathing, do not stop while a being is alive, and neither does attention beholden to life. Our embodied existence and the vitality of plants are incessantly attentive to the basic conditions for living, above all, those contained in the elements.

If we wish to discover vegetal mindfulness in ourselves, we should look no further than our bodies' unconscious attention to the surroundings. Like the leaves of plants, our skin senses humidity and temperature, light gradients and vibrations. Without knowing it, we pay attention to the world on the surface, from the porous, essentially open, cutaneous membrane that keeps our living bodies intact and, at the same time, communicates with whatever lies beyond them. Like leaves, we breathe on the surface of our skin, not only through our lungs. Every one of these innumerable tiny breaths is a channel between the body and the elemental milieu wherein it is immersed. Seemingly effortless, the breathing of the skin and of the leaf, their minute attention to air, is nourished by the living energy of life. We cannot (and should not) control this process, but we can shuttle between the breathing of the skin and of the lungs, between our unconscious and conscious attention to the world, and between vegetal and human mindfulness within and outside us.

The plants' rootedness in the soil determines their attentiveness to the places they inhabit, the places they cannot leave behind or negate in the manner of animals or humans. Being with plants, we learn to consider the fine grains of a place, which is much more than an isolated point on the system of spatial coordinates or a site through which we momentarily pass. Living attention to the location of growth is inscribed into the spatial pattern of vegetal bodies: more branches on the sunnier side of the tree, less roots if a plant grows in close proximity to another of the same species, and so forth. By the very way plants inhabit their environments, they direct us toward the elements, to which they attend. For me, this idea will remain forever associated with a singular place: the wooded setting of McMichael Canadian Art Collection in Kleinburg, Ontario, housing the works of the Group of Seven and the Canadian First Nations. The northern pines of the Kortright Centre for Conservation, where the gallery is located, much like the trees depicted in the paintings of Fred Varley or Franklin Carmichael, express an evocative interplay of vegetal and artistic attentions to places and to the elements.

Over several years I spent almost every weekend at McMichael and the vicinity, alternating between the works of art and the woods that encircled the gallery building. Each visit changed not only my experience of that place but also of time. After awhile I could no longer tell if I spent twenty minutes or two hours with plants. My sense was that everything happened more slowly and that in that deceleration, which facilitated a closer attention to plants and to the elements, living became more vibrant. Time and again, time was gained for life.

26 January–2 February 2014

8

FEELING NOSTALGIA FOR
A HUMAN COMPANION

The companionship that the elements and plants provide is faithful but has its limitations: I cannot address such companions by speaking, except poetically and nonreciprocally. Human companionship is less steady, more risky, but it can translate into an exchange of words that flow from the one to the other and back again. Is it for this creation of a common world in speech that we long when we experience nostalgia for a human companion?

As I have observed before, the sense of being alone *in* nature is somewhat skewed in the absence of a feeling that we are alone *with* nature. Having in mind Rousseau's solitude during his "happy reveries" among greenery, flowers, and birds, I asked: "Is it the case that we are—still or already—immersed in solitude when we are with animals or plants? What does being with these nonhuman beings mean? Doesn't being *in* nature, as we say in everyday language, ineluctably create a broad transhuman community: being *with* nature? If so, then what is our place, and that of our thought, within such a community, and where are we situated with respect to the place of plants?"[1]

To be *with* nature, we must first take an apprenticeship in its nonverbal languages. I am not referring to something like the reading of signs— for the arrival of spring when the first fragile flowers emerge from the earth, or for the imminent rain when clouds gather menacingly overhead. What I have in mind is the way living beings express themselves in spatial configurations and along their lifetime, in conformity with the seasons and the places they inhabit. Cherry blossoms are the tree's springtime expression; the ripe cherries are its summer manifestation; the changing leaves are appropriate to the autumn; and the bare branches covered with dormant buds are the face the tree turns toward the winter. Changing throughout the year, the cherry tree does not express anything other than its own being over time. I could say that I am following Leibniz's philosophy here, except that, for him, each is ultimately an expression of God, or of divine substance, whereas I underline the diversity of living self-expressions, irreducible to a single magisterial source.

So, being with nature implies a heightened receptivity and openness to the endless variety of nonverbal languages that surround us in the garden, in the field, in the woods . . . In our houses and apartments, chairs, tables, stoves, bookshelves, etc., also express something, but what they stand for does not go beyond our needs. Unlike the smallest blade of grass, these manufactured objects do not express *themselves*—that is the offshoot of Aristotle's distinction between things that are according to nature (*kata phusin*) and cultural artifacts, the objects of *technē*. Disturbing this neat differentiation, a work of art attains some degree of self-manifestation on the hither side of need, though even it does not do so as fully as a marigold growing in a garden. Spending time *with* nature, instead of simply *in* it, we let our expressiveness resonate with that of the animals, plants, and even minerals or rock formations we encounter. The difficulty, however, is that the fullest human self-expression relies on words, addressed to the other. Does this suggest that our resonance with the natural world will be deficient, if it is not at the same time a coresonance with another human?

Despite the richness of nonverbal languages offering to all our senses a tremendous variety of sights, sounds, textures, and smells, something is missing in the midst of this plentitude. The second creation narrative in the book of *Genesis* conveys the loneliness a human feels when surrounded by vegetal nature. The original divine judgment on something that is not good concerns this solitude and occasions the creation of birds and animals from the same material—the earth, *adama*, in the feminine—from which Adam was formed (Genesis 2:18–19). From the act of naming these living beings, human language is born, but it remains deficient and nearly useless without anyone to share words with, and not only call by name. Such is the pretext for the birth of Eve from Adam's side (*tzela'*). Even if it seems that this second story of human origins is less egalitarian than the first, where the man and the woman arise together, side-by-side, it nevertheless holds a hidden promise. Here Eve comes to rescue Adam from his hopeless solitude, and so she is his redeemer. Poorly translated as "helper," the word *'ezer* (Genesis 2:20), which describes the role of Eve, connotes redemption and deliverance of the kind extended to us by God himself. (For instance, in Exodus 18:4 the explanation for the name Eli'ezer is that "the God [*elohei*] of my father was my help ['*ezri*], and delivered me from the sword of Pharaoh.") The first woman, not counting Lilith, was the first messiah. Adam's nostalgia for a human companion is, thus, resolved by his deliverance by the other who (though, presumably, crafted from his flesh) is different from him, who can address herself to him and be addressed in words, and who can, together with him, remark on the wonderful variety of plants and animals that populate the Garden of Eden. Everything already created by God is re-created in the world that issues from speech.

Plants, to be sure, respond to our care without saying anything. They do so by expressing themselves more exuberantly, by spreading more branches, unfolding more leaves, or opening more blossoms. They exist, they *are*, more intensely, and this intensity is of a piece with their extending themselves further into space, or, in a word, their growth. On a certain level we may experience a similar bodily welcoming of existence, the same opening unto the world, as a plant. But our, human, openness will be incomplete unless it takes place with another human being or with other human beings. A healthy vital rhythm is that of retreat and reemergence, and it is when the first phase of taking refuge—for instance, in the vegetal world or, for some, in the world of books—becomes too prolonged that we feel the desire for having an experience in common with another human. That is why the beauty of a wildflower or of a sunset might not be fully appreciated without being coexperienced with the other.

To put this in slightly more formal terms, what we miss when we are immersed in an experience of (and with) plants and of (and with) the rest of nature is the very inwardness or interiority that we seek in another human being. In reflecting on the kind of shelter the vegetal world granted to us, I mentioned how it destabilized the classical metaphysical distinction between the inner and the outer. The plants' self-expression, in all its marvelously extended character, seems to veer on the side of exteriority. Perhaps, then, one of the sources for the nostalgia we feel is that we sense a deficit of interiority in the vegetal world. If that is so, then our nostalgia is still nourished by metaphysics, which refuses to acknowledge, on the one hand, the unique subjectivity of plants and, on the other, the noninterior, extended facets of a human subjectivity. That is what is missing when everything is given and when everything generously and persistently gives itself to all our senses, as plants do.

Traces of nostalgia for a human companion are detectable even in those literary narratives, such as Antoine de Saint-Exupéry's *The Little Prince*, where a flower appears as the singular recipient of affection.[2] There, the rose adored by the little prince reciprocates as a human would, making her own demands and professing her love. In other words, the flower is symbolically endowed with interiority, not to mention with a feminine identity in relation to the male lover. The same happens in Novalis's novel *Heinrich von Ofterdingen* and its mysterious *blaue Blume*, or blue flower, with which the main character falls in love and which is later revealed to be a substitute for his beloved, Matilda.[3] Often a persistent nostalgia for the human does not allow us to see flowers as nothing but flowers; instead, we wrap them—and other plants that matter to us—in layer after layer of symbolic significance, cultural meanings, and utility. It is not by chance that Freud puts "the language of flowers" at the behest of human psychic life wherein it accomplishes a displacement of the sexual drive. Indeed, culturally, flowers are usually assigned the task of mediation between romantic partners, but to narrow their language down to this function is to impoverish their self-expression.

Imagine that the love of plants (or what I like to call *phytophilia*) is genuine, that is to say, that it is not a mere fetishistic diversion of the drive to a substitute for another human. The surprising thing is that, if it excludes other humans and "simple" plants, this otherwise admirable attitude may turn out to be unjust! We cannot avoid privileging the singular being or beings we love over those that do not evoke affection in us. With phytophilia as our Ariadne's thread, we strive to protect, nurture, and care for the singular plants we encounter, while forgetting about those that do not attract us. Thus, in Saint-Exupéry's small book, a boy is so enamored by a rare flower growing on his planet that he disregards all the other plants and blossoms. "I am responsible for this flower," he says, trying to prevent a sheep from eating it.[4] But he does not care about the common "very simple flowers . . . with a single ring of petals, occupying little space and no trouble to anybody."[5] And in a desert he passes over the only flower he meets, "a flower with very few petals, a flower of no importance . . . "[6] In the climate of general indifference to plants, a phytophile runs the risk of retreating into a private utopia that is powerless when it comes to preserving and protecting the increasingly fragile vegetal life from economic practices that convert it into a set of commodities. There is nothing inherently wrong with phytophilia, but we must strive to impart the love of plants to as many humans as possible so that the singularity of their individual affections could amount to a universal care for vegetal life. At its finest, the nostalgia for a human companion is a longing for justice.

As a result, phytophilia is not tantamount either to misanthropy or to the hatred of animals. In a gross misrepresentation, some proponents of "human exceptionalism" have characterized me as a "green misanthrope,"[7] while "animal rights" advocates have insisted that my defense of plants harms animals.[8] But there is nothing to suggest that attention to vegetal self-expression and care for plant life are incompatible with attention to and care for other modes of living. In fact, regardless of how indispensible our singular encounters with plants are, a shift in our cultural and, later on, agri-cultural and political behaviors toward them will happen only once we share our love of plants with others. Before this comes to pass, each of us must first rediscover, in solitude with nature, an experience of plants that has not been subject to the mutilation, trivialization, and instrumentalization to which it has been submitted, at least in our Western culture. Like everything else, nostalgia for a human companion in relation to plants has its appropriate time or season: we should not succumb to it prematurely, before we have reexperienced our natural environment differently. If a cherry tree blooms before its time, its flowers will be bitten by the nightly frost; if we approach another human too early, we are likely to sense the cold of misunderstanding and, at times, ridicule. How to let grow and ripen a human community around the vegetal world—that is still an open question.

24 March–2 April 2014

9

RISKING TO GO BACK
AMONG HUMANS

How to let grow and ripen a human community around the vegetal world? And, first of all, how to achieve one's own appearance among humans? Let us begin with plants, once again.

From the smallest clover to an oak tree, plants appear to the light of day thanks to their germination and growth. While still in the dark and moist density of the soil, a seed begins to develop *in and as the middle* by extending itself both up and down, by moving between the polarities of the rootlets and the shoots until its upper portion emerges to daylight. This emergence is not just one among many examples of phenomenality or of how beings show themselves. Rather, plant germination and growth prefigure other kinds of coming-to-appearance by means of self-exposure to the other or to others, be it sunlight, water, the sentience of insects, or human vision and sense of smell. Plants are the events of finite being. Living *phenomena* are *ta phuomena*, the growing beings Aristotle invokes in *De Anima* (413a, 24–26). Their—and our—appearing is never complete and therefore cannot be arrested in the form of an immutable truth. From the middle, the milieu where it is at any given moment, it can only try to respond better to the changing context, elements, and conditions of life without putting forth a one-size-fits-all organic shape or solution.

The other reason for the essential incompleteness of the living coming-to-appearance is that something of the growing being is kept hidden, both physically, as the root is in the soil, and at a different level, as life itself, encrypted in everything that seems to expose itself entirely and unreservedly. It is always possible to retreat to this clandestine reserve, in the manner of perennial flowers dormant in their bulbs during the winter. *Ta phuomena* make an appearance with the view to their disappearance, and they preserve their nonappearance (which has nothing to do with metaphysics or with noumena) throughout their appearing.

Last, the phenomenality of "growing beings" is incomplete because they need each other and the elements in order to embark on their lifelong joint coming-to-appearance. Their appearing in the world is a coappearing together with everything that supports their growth. Taken by itself, a phenomenon, as much as a *phuomenon*, is insufficient (even for itself!) and nonviable.

Lest we forget, humans are also growing beings and, as such *phuomena,* we have no choice save for growing together with others, including the elements, plants, animals, and our fellow humans. That is not to say that we lack the reserve of nonappearance; if anything, this reserve has been overcultivated in our civilization, which idolizes the individual dwelling, the private sphere, and psychic interiority. These structures, moreover, have been erected as barriers, supposed to isolate the self from its others, or, if you will, as shells ensuring the integrity of our innermost kernel. Coming-to-appearance is then understood as a quick emergence from this shell in order to snatch something from the outside world—from the elements, plants, animals, and human others—and to hoard this booty in the storehouse of one's experience, home, or bank account. Nietzsche expressed this movement well in the introduction to *The Genealogy of Morality*, with its indelible image of ourselves as "winged animals and honey-gatherers of Spirit," yearning for "'bringing home' something."[1]

If we are willing to learn coming-to-appearance from plants, we must go back among humans differently. We must, more precisely, cherish our incomplete coappearances with others, not positing as a goal the appropriation of what doesn't belong to us, but turning our growing-together into a desideratum in itself. We should give up neither on our exposure to the social world nor on the hidden reserve of being alone in and with nature. And we should embrace the kind of appearing that does not find its value and sense in the fabrications of "objective truth" but in truth as a betrothal, a faithfulness to everything that and everyone who makes it possible.

Risking to return among humans, I plunge into the different ways in which one can make oneself public, from a simple physical self-manifestation "in person" before others to giving an academic presentation or publishing my reflections, such as those you are now reading. Why are these ventures risky? For one, because they inexorably entail a coappearance with others who are outside my control and who might not be as keen to learn the phenomenality of growing beings from plants. The dangers range from misunderstanding to an outright dismissal and trivialization of our unbreakable connection to the vegetal world. Instead of growing together, most people tend to undermine and uproot what they are rooted in, that is to say, their human others, plants, and the world of the elements. (This, by the way, can be a more encompassing characterization of our sad environmental and social predicament today.) In response to such aggression, one is tempted to protect oneself by growing thorns, as certain plants do. And the stakes are raised even more once we realize that the judgment "it is not yet time to go among humans" might cause us to miss the opportunity of rejoining them altogether and so leave us isolated in the experiencing of and with plants.

On the one hand, in the Western tradition, from Plato to Hannah Arendt, the meaning of achieving one's own appearance depended on the difference between the event of biological birth and a rebirth through philosophical or political action. The two events—and with them the realms of nature and culture—seemed to be discontinuous largely due to the animalist prejudice of the philosophers themselves, who placed undue emphasis on absolute separation at birth. On the other hand, the original philosophy, elaborated by Luce Irigaray, has provided us with the tools necessary to mend this artificial divide by deriving civic and other kinds of identity, not to mention the very humanity of the human, through the cultivation of sexuate differences. No longer do thought and action have to assert their right by setting themselves over and against the world into which we are born; rather, they can grow and mature by sharing this world.

With vegetal life in mind, we might ask, in turn: What if the phenomenon or the *phuomenon* of plant germination were behind both events of birth and rebirth? If that were so, then there wouldn't be a traumatic break with what or who gave us life, but a continuous appearing, forever indebted before and relying on its "soil." Although the umbilical cord is cut, we are rooted in the other as much as in ourselves, and to go back among humans and among plants is to return to a milieu we have never really left.

In Arendt's philosophical universe there is no intermediary between biological birth and action, between the tragic and mute loneliness of the separated (born) subject and the solitude that allows it to hear itself speak and therefore to think. She does not admit the sense of being with nature in a silence that transcends the narrow confines of both experiences and puts us in touch with the existence of plants. Venturing a comeback among humans is breaking this silence of growth and metamorphosis, which our bodies undergo in parallel with those of plants. Most often, when we appear before others, we do so by exposing our physical extension before their gazes as well as by speaking—hence, by denuding our subjective interiority. This disruption of silent life is irreversible; as Samuel Beckett poignantly put it, "Silence once broken will never be whole again."[2] "Is there then no hope?" Beckett goes on to ask on the same page. There is, I would retort, so long as the uttering of words does not cut us off from the vegetal world. So long, that is, as the uttered words respect and preserve the silent existence of plants and our equally wordless experience with them.

But respectful words alone fail to mitigate the risk of going back among humans. In the desire to communicate with them, my incommunicable involvement with plants is lost and something of myself is missing. It is not even assured that when I say "plant" the others understand what (or who) I am referring to, namely a growing, appearing, infinitely generous, living being. They may, instead, think of foodstuffs or construction materials, photosynthesizing biological machines or green self-reproducing matter, and so forth. It is not that the referent is lost behind the word, but that the enduring growth and metamorphosis of plants are blocked or obstructed in the spoken (or written) medium itself.

And what if one coappeared with and before others in environments where plants grow: botanical gardens, forests, or parks, for example? In these places, words reveal their insignificance in comparison to the vegetal life that surrounds us; they melt back into the silence of plant existence, around which a transitory human community can start to ripen. Whenever I am invited to give a public talk on the subject of my "vegetal research," I secretly hope that it might take place in the presence of both humans and plants. I want my words and my writing to circle back to trees and flowers, the shrubs and the grass, even if they are addressed, prima facie, to fellow humans. Thought, speech, and writing should be not only biodegradable, as Jacques Derrida wished—that is: they should not merely dissolve into the soil of culture, turning into fertile compost—but they should also germinate and give off new shoots, in the first instance, by letting plants grow.

Publishing conventions did not permit me to embed the seeds of flowers in the cover of *Plant-Thinking*, which I envisioned buried by its readers after they flipped the final page, so that the book would literally blossom. But I was very pleased to see this idea realized by the organizers of the 2013 Annual Meeting of the Society of Plant Signaling and Behavior in Vancouver, Canada, who produced conference notepads with a mix of perennial and annual flower seeds included in the front and back covers. The notebook is now blooming in my garden. Although this is a very literal example, it shows how one type of public appearance before others can revert back to the plants themselves.

In sum, the greatest risks of going back among humans, in a vast variety of manners available to us, are drowning the silence of vegetal life in the empty chatter of speech; forgetting that our own coming-to-appearance has much in common with and is indebted before the germination, growth, and exposure of plants; and neglecting to give plants their due. Regardless of all the precautions I may take, these three dangers seem unavoidable, unless I persist in asking nature—*phusis,* the ensemble of creaturely growth—for a solution.

9–22 April 2014

10

LOSING ONESELF AND ASKING NATURE
FOR HELP AGAIN

The theoretical quarrel between Kant and Hegel, which, to a fair degree, is still molding the philosophical landscape today, can be summed up in the question of rootedness. Kant's ideal of autonomy suggests that a human should be rooted only in herself or himself as the subject of reason independent of all capriciousness and contingency. Hegel, for his part, acknowledges the rooting of every one of us in another human being (or in other human beings), in whom we finally discover our inter-subjective identity, albeit through the negation of our "immediacy."

Surveyed superficially, the two German thinkers offer completely different perspectives on the sense of losing and finding oneself, assuming that Kant turns inward, toward the principle of transcendental rationality, whereas Hegel turns outward, toward another human in the search for a "truer," more authentic, richer self. But aren't the two, for all intents and purposes, beholden to the interiority of the human? Isn't the only variety of growth they can tolerate inward? Insofar as radical, non-human exteriority announces itself on their horizons at all, it is nothing more than the unknowable noumenal realm (Kant) or "raw" nature that must be mediated into concepts through painstaking negation (Hegel). Reason and Spirit are ingrown entities that, within certain limits, absorb what lies outside.

To the extent that it has let anything grow at all, our metaphysical tradition has sanctioned nothing other than inward growth and, therefore, a growth that is already stunted and that fails to learn from plants. It has shunned the outgrowths or excrescences that plants are and substituted for them a spiritual or cultural development predicated on "deepening" our inner resources. Although this option is viable for the rest of *phusis* as well, it maintains something of an exceptional, if not anomalous, character, especially when it is presented as the sole path open before us. For a living body, to grow inward is quite pathological (just think of the pain caused by ingrown nails), but it is this pathology that has come to constitute the norm of psychic life over millennia in the West. Whether one is ideally rooted only in oneself as an autonomous subject or in another human to the exclusion of other modes of life, one develops, by and large, as an ingrown nail does, causing tension, inflammation, disruption, and disease in one's milieu. Figuratively speaking, instead of living, one festers, and the sad outcomes of this planet-wide transgenerational collective festering are observable in the environmental crisis that has gripped our world.

We lose ourselves when the choice we face is between retreating into the prisonhouse of loneliness or going back among humans whose company we mistake for radical exteriority. Paradoxically, in order to recover ourselves we must lose ourselves better by learning how to grow *outward*, to be an excrescence that, while remaining rooted, knows how to grow *with* nonhuman others—the elements, plants, and animals. The best place for undertaking this apprenticeship is the forest, which in the West as much as in the East culture as a whole, with its predilection for the vast expanses of cultivated fields and concrete blocks of cities, tends to destroy. Heidegger still had enough sense to situate the human opening in the forest, in the *Lichtung*, where the open with its light is limited, shadowed, sheltered, and ultimately preserved by the trees that surround it. Worldwide deforestation, on the contrary, signals the irretrievable loss of exteriority, of limits and boundaries, of sheltered light, and

of life itself on a planet that is becoming increasingly cultivated, domesticated, and eroded. If this trend continues unabated, soon enough we
will no longer find ourselves surrounded by forests as humans have
been throughout their history. On a deforested earth, the nihilistic
dream of the European Enlightenment would finally be fulfilled: totally
inundated by light, unfiltered through the canopies of trees, everything
would become visible, transparent, observable without obstacles from
the air, and devoid of shadows. Hence, the possibility for any kind of
vision would vanish.

It will not come as a surprise, then, that the idea for *Plant-Thinking*
was born in a beautiful forest of Gerês, the region that spans the border
between Northern Portugal and Spain. When, in December 2008, I came
to this enchanted and enchanting place, I still did not have as much as
an inkling that I would, in some sense, find myself there by losing myself
better, more thoroughly. I felt that Gerês, perhaps more than any other
forest I had been to, was not only a retreat but a place for breathing,
existing, seeing, imagining, and thinking anew. Besides sheltering various clearings, where wild horses abide, among its moss-covered trees the
forest itself was a clearing, a *Lichtung*—at least, in the way I perceived it
and perceived myself in it. It was there that my thinking first grew outward, toward the plant (and animal) life that surrounded me, rather than
feeding on itself, that is to say, on the tradition of Western philosophy
and the decay of traditional metaphysics that marks our contemporary
situation. As it grew outward, together with and following the lead of
plants, my thinking gradually came to view thinking as such in terms
of an excrescence, an ex-tending intentionality, or an outgrowth of the
Aristotelian *to threptikon*. (Tellingly, one of the books I took with me there
was a masterful commentary on Aristotle by Claudia Baracchi.)[1] Would
this have happened without my own exposure and striving to vegetal
exteriority? In retrospect, I can only speculate on the exact cause, but it
is clear that spending time in Gerês was a catalyst for the timid departure
of my thinking from its ingrown condition.

I have just noted that it was necessary to "los[e] myself better, more thoroughly." The expression is not a vain turn of phrase but a hint as to how nature can help us overcome our desperate dialectical vacillation between losing and refinding ourselves. How so?

In order to lose oneself, one must first lay claim to and become a proprietor of oneself. But what if that proprietary claim were, as I've already admitted in chapter 2, an extravagant fiction? It certainly does not hold for plants and it obstructs the vitality of human life too. In the text on "the plant that is not one," I've expressed this positive side of vegetal non-identity as follows: "The plant that is not one grows with the other only because it does not grow with itself; being with the other is non-transcendentally possible for a vegetal being that is not, in any case, with itself. Growing-with is the expression of growing-without, an infinite chain of supplementations to the absence of vegetal unity and identity. As such, it is to be rigorously distinguished from growing-together and, especially, from the notion of co-evolution—e.g., of honeybees and the clover flowers they pollinate—where new capacities for adaptation are mutually shaped thanks to the ongoing contacts of distinct species. If growing-with really expresses growing-without, dissemination, and being-not-one of and at the origin, then the conjunction of growths comprising, on the one hand, what appears to be a single, discrete being and, on the other, a community of beings must be understood in terms of its inherent falling apart into a non-totalizable multiplicity. There is no growing-with without falling apart: such is the axiom of vegetal democracy."[2]

Our image of subjective wholeness is as illusory as that of a simply divided subjectivity, split between itself and the human other, the inside and the outside. The common supposition undergirding both views is that the subject is ingrown, having developed either exclusively within itself or in relation to the human other. But this supposition ignores (to put it in somewhat dry, academic terms) the environmental construction of subjectivity, namely the nonhuman world of elements and plants that continually bestow upon us the necessities and pleasures of life. Growing with this world, as plants do, we are the excrescences or outgrowths that blossom not only when we cultivate our sublime, spiritual interiority but also when our psyches, inseparable from our living-breathing bodies, are extended, striving to an exteriority no longer exclusively human. The question is not so much "what or whom do I find when I look for myself?" but "*where* do I find myself (at my desk, in the forest, in the garden, in the mountains . . .)?" And "*with* whom or *with* what?"

In the wake of this last question, it is imperative to add that, at the same time that we are outgrowths in our living and thinking practices, we are the with-growths that, whether we admit it or not, flourish with the rest of our environment. Such "growing-with," which echoes the condition of being with nature I described in chapter 8, ought to be heard and sensed with all its existential overtones, as the ignored backdrop for the development of the Heideggerian being-with, *Mitsein*. The bond of the *with* is not cemented by a common essence, but emerges from shared life, understood based on its vegetal determination as growth. It is, in part, this absence of essential ties that explains why growing-with is already a falling-apart, whereby those who forge a community of growing beings do not suffocate in the rigid schemes of *essentia* or *quidditas*—in a word, in the deadening order of whatness.

The other reason for the disjunction at the heart of growth is that plants are not organisms, that is, they are not living *totalities* with organs subordinated to the demands of the whole. In growing with each other, their parts preserve their independence and so fall apart. Our human with-growth should be thought of along the same lines if it is to avoid the trap of a fusion with the whole in which each is dissolved (almost) without a remainder. This is, probably, the most important assistance vegetal nature can provide us with, placing us on the hither side of the dialectics of loss and recovery.

24 April–5 May 2014

11

ENCOUNTERING ANOTHER
HUMAN IN THE WOODS

On those winter or spring days when fog from the Atlantic rolls in thick patches among the pines, the woods appear to be entirely transformed. On one such occasion I walked through the usual path, but there was suddenly nothing familiar about it; I could barely discern the shapes of the trees, which were actually very close to me. The space of the woods felt different. More than ever, this space felt as one of proximity rather than of remoteness, especially because I no longer relied on the "distance sense" of vision to find my way, but, instead followed the path intuitively or, at best, used the sense of touch. I thought that Henry Thoreau's description of a barred owl was very pertinent to this experience: "Thus, guided amid the pine boughs rather by a delicate sense of their neighborhood than by sight, feeling his twilight way."[1]

Besides sights, sounds were also muted and withdrawn. The silence of the woods was soft and not at all oppressive. The air was so saturated with ocean moisture that one could observe its every movement, which occluded the outlines of beings—vegetal, animal, and possibly human. Did the pine forest teach me a lesson in unlearning the taken-for-granted images, habitually supplied by my senses? If so, then it did a better job than all the theories of philosophical skepticism and all phenomenological reductions combined. Shrouded in mist, with which it also covered me, it returned me to a state before "sense certainty," that is to say, to the indeterminacy of existence before it lends itself to self-assured judgments and interpretations.

I am also conversant with the lucidity of the pine forest, which was poetically recorded by Francis Ponge on the morning of August 13, 1940 (an ominous date, still very much at the beginning of the war, which contrasts so sharply with the calm of the woods).[2] The ease of passage and of reflection among the pines, however, depends on the elements, from which the vegetal world cannot be entirely dissociated. The day in late March, when dense fog transfigured the woods, was very different from Ponge's August experience. It is not that lucidity vanished; rather, it was altered, together with the appearance of the site. It became obvious to me that the forest was not only a "site," much less a "landscape," but the very embodiment of a place, which admits everything and everyone into itself. When Plato contemplated the place in terms of *khôra* (usually translated as "receptacle") in *Timaeus*, he must have been—or, at least, dreamed of being—in the woods. Surrounded by the barely visible pine trees and enveloped in fog, I felt especially sheltered in the place that received me so. And I imagined that I was alone in it.

But, by their nature, places are not exclusive. As I have already written, they admit everything and everyone into themselves, so much so that they are minimally defined by this welcome. Whoever (or whatever) is in a place is hosted there, without ever merging with "the receptacle." In a place, there are at least two: the being that occupies it and the place itself. And, if there is no place corresponding exclusively to a unique being it contains, it is also because places are coinhabited—either simultaneously or at different times. The logic of possessive appropriation, which commences with animals rather than with plants, does not accept this (for a lack of a better term) axiom of placeness: it converts places into territories to be marked and controlled as one's own. My imagining that I was alone in the misty woods unwittingly harked back to something of this logic, though, most likely, I received this impression from the uncommon silence that surrounded me and the sense that space itself had shrunk and that the woods had closed in over and around me, right next to my skin.

Everything is, no doubt, simpler when one finds oneself (or deems oneself to be) face-to-face with the natural world, in the absence of any other humans. But this absence is never complete. Even when I am immersed in my thoughts, I rely on words that belong to the generations of countless others. Although, similar to Jean-Jacques Rousseau, the relation to plants "constantly reminds me of the meadows, the waters, the woods, the solitude, above all the peace and tranquility one finds in the midst of all these things,"[3] this solitude is highly questionable, because I am not really alone with plants and I do not fully leave human company when I am in their midst.

As, in my presumed solitude, I was reflecting on this state, I discerned what appeared to be a human silhouette among the trees. It was difficult to be certain at first: there was but a hint of movement among the shadows of tree trunks in the thick fog, which concealed the difference between two vertical beings, the vegetal and the human. The movement was very slight, probably because the other human being (if it was one) stopped in order to decide on the best path to take, given poor visibility. In this moment of suspension and suspense, when the promise of an encounter was still vague, it occurred to me that I was alone and not alone at that precise time in the woods, as well as everywhere and always. Among the trees and the shrubs, I was in the presence of other living beings who existed otherwise than I did and who included not only plants but also many insects and other animals, fungi, and possibly humans. Outside the space or the place of the woods, I can confine myself in total solitude to my room or to a study (as Descartes did, sitting by the fireplace in which he burned and tried to annihilate without remainder traces of vegetal nature, along with the traces of all "extended" reality), but I will nevertheless retain the sociality of language, the memories and anticipations of encounters with other living beings, and the invisible bond that connects me to the vegetal world through breathable air.

The feeling of being alone and not alone is not tantamount to Kant's "asocial sociality." It has nothing to do with a careful balance between the "spheres" of individual subjective existence and dependence on others. Instead, it goes back to the idea of falling apart in growing-with, which I discussed in the previous chapter. This experience preexists our conceptions of sociality and individuality, which is why I call it a "feeling" or an "intuition." An alternative to solipsism and to the "oceanic feeling" of merging with the world, this intermediate condition participates in the movement of growth. With and without—with *as* without—the other.

Being alone and not alone is what I felt very distinctly in the fog that enveloped me, the trees, and perhaps another human being. Did I get in touch with something of the background for existence that allows those who exist to be themselves and to be with others? After a moment when all was stopped, which seemed to last for a long time, movement resumed, and I realized that it was, in effect, another human being walking through the pine forest. Everything proceeded at a quicker pace from then on, seeing that the instant of realization happened quite late, when the two of us were about to cross paths under the curtain of dense fog. As we walked by one another—was there a sense of this "we" in that moment?—the silence of the woods remained undisturbed. Neither of us said anything to the other, but I nodded as a sign of acknowledgment for a fellow walker in the woods and received what I interpreted as a nod in response.

Was that an encounter? There was not a single word exchanged between us, and each had barely recognized the other as human before we nearly ran into one another. But how else could an encounter in the woods—which is also, at the same time, an encounter *with* the woods—come to pass, if not in silence and with the minimum of recognition? What could be more faithful to the sense of being alone and not alone?

How could an encounter happen, be a happening or an event, if not on the grounds of indeterminacy and of a singular occurrence, that is, of not having a future? It needn't be groundbreaking or traumatic, just as the event of plant growth is none of these things. An eventful encounter, such as the one I am describing here, can be mundane, so much so that it may elude our attention, pass unnoticed as an encounter, or become a part of the past before being re-presented. But it is in this inconspicuousness of the mundane, its nonarticulation in speech, that something extremely significant may hide—for instance, the realization that I am both alone and not alone.

Was that another human? The question is not as strange as it may sound. Of course, there was a period of indecision, when I could not tell whether I was noticing a human figure among the trees or whether I was imagining something that was not truly there. This is not the uncertainty I have in mind, because it dissipated as soon as our trajectories further approximated one another and crossed. So the issue is not foregrounding the human *figure* from a more or less fuzzy visual background, but detecting the human as a human *being*. For Descartes—to return to him once more—the problem of identifying a human was a major source of worry, because, in the opinion of the French philosopher, it was within the realm of the possible that the beings he saw around him were humanlike robots ("But what do I see aside from hats and clothes, which could conceal automata?").[4] For that reason, Descartes felt utterly alone, as he was not really capable of encountering another human without suspecting that it was not one.

But exactly why would robots be drawn to the woods? What would prompt them to stroll around in vegetal nature, even—and especially—in inclement weather conditions? Wouldn't such behavior be interpreted as useless and irrational within the technologist paradigm they exemplify? Despite passing each other in silence, the other person and I have encountered one another thanks to the woods where the encounter took place.

Did each become a little more human as a result? There was an implicit solidarity in that brief passage, not in the form of direct allegiance to one another, but in the form of commitment to the vegetal world. The silence of the encounter echoed that of plant life; it was a sign of respect for plants and for each other. The two silences resonated with one another only because they resonated with vegetal silence. Could it be, then, that we are least robotlike when we encounter the natural world and meet other humans there?

Having questioned, at a relatively high speed, the meaning of the encounter and that of the human, I will not extend the questioning impulse to the final substantive term in the title of this chapter, namely to the woods. For too long, philosophers, including Descartes and Husserl, have doubted and reduced (mentally eliminated) external reality and especially the natural environment. I leave it up to them to debate whether those were really woods, whether the entire thing was a dream, to what extent I could trust my senses in the dense fog that enveloped everything, and so forth. The woods, the human relation to them, and life itself precede all matters of critique: neither radical doubt nor reduction is plausible unless the woods have already energized life and thought. It remains to be seen what shape this energy assumes, what I can do with it, and what it can do with me.

10 May–12 June 2014

12

WONDERING HOW TO CULTIVATE OUR LIVING ENERGY

In the West the term *energy* is marked with the force of deadly negativity. It is presumed that energy must be extracted, with the greatest degree of violence, by destroying whatever or whoever temporarily contains it. More often than not, it is procured by burning its "source," in the first instance, plants and parts of plants, whether they have been chopped down yesterday or have been dead for millions of years, the timescale sufficient for them to be transformed into coals or oil. Without giving it much thought, one supposes that the only way to obtain energy, whether for external heating or for giving the body enough of that other heat (namely, "caloric intake") necessary for life, is by destroying the integrity of something or someone else. Life becomes the privilege of the survivors, who celebrate their Pyrrhic victory over the ashes of past and present vegetation and other forms of life they commit to fire.

Seeing that, for Aristotle (who still maintains a strong hold on *energeia*, a word that he introduced into philosophical vocabulary), the prototype of matter is *hylē*, or wood, the violent extraction of energy paints a vivid image of the relation between matter and spirit prevalent in the West. A flaming spirit sets itself to work by destroying its other and triumphs over the wooden matter it incinerates. The price for the energy released in the process of combustion is the reduction of what is burned to the ground. And, unfortunately, the madness of metaphysical spirit, which sets everything on its path aflame, tends to intensify. As I have written elsewhere: "The tragedy of the twenty-first century is that we have taken it upon ourselves to . . . to burn everything that is combustible, including, at some level, ourselves. It is incontrovertible that the extraction of energy by burning things predates not only our epoch but industrial modernity as well. But there is a tremendous difference between the fires kindled with a few branches in a prehistoric cave, the coal burnt in the English factories of the eighteenth century, and the contemporary combustion of biofuels. The early representatives of humankind set a small bit of the present or immediate past alight; rapidly industrializing Europe submitted the deep past of vegetal and animal life to fire; today's blaze bears a total character, where the past, the present, and the future burn together in a process that is indistinguishable from the incinerated life, from its production and reproduction for the sake of being consumed as energy."[1]

It is not that plants are exempt from the general combustibility that, for Schelling, defined the very living of life.[2] They release oxygen and so provide the elemental conditions for the burning of fire. But the vegetal mode of obtaining energy—especially that of the solar variety—is nonextractive and nondestructive; the plant receives its energy by tending, by extending itself toward the inaccessible other, with which it does not interfere. That is still another vegetal lesson to be learned: how to energize oneself, following the plants, without annihilating the sources of our vitality.

Besides extraction and destruction, the Western paradigm of energy includes a seemingly more positive dimension: production. In fact, Aristotle's *energeia* literally means setting-to-work, activating, en-acting (and the Latin rendition of this Greek word is, precisely, *actus* or *actualitas*). To his credit, Aristotle thought that plants also put themselves to work and actualized themselves by growing and reproducing. In this, the Aristotelian considerations of vegetal life are much more perceptive than those of the moderns, who deem plants to be immobile, purely passive, and therefore fit to serve as materials for energy production. This is not a minor point: behind the so-called subject-object split of modernity is a more profound divide between the energizing and the energized, the storehouse of energy and those who can tap into it at any moment.

One could argue, more or less accurately, that the main flaw in Aristotle's notion of energy is its unqualified allegiance to a productivist worldview. By definition, there is no energy or life without work (*ergon*), without the setting-to-work as a restless process that, in advance, finds respite only in the finished product. The Greek philosopher proposes to his readers something that, today, appears to be a truism: there is no fulfillment but at the end. The fruit is the end of the plant and, therefore, the fulfillment of its life. Ejaculation is the orgasmic end (on the part of the male, mind you) of the sexual act and, therefore, the fulfillment of desire. But what about the energy that is not at work and that, without energizing some end product, finds fulfillment in itself or in flourishing between me and the others, whether human or not? Isn't my being with plants, outside the need to consume them, an instance of fulfillment without an end? Kantian and post-Kantian thought typically identifies this condition with the aesthetic sphere. But its scope is incalculably broader than that: it befits finite existence, life itself, which is not fulfilled upon its physical termination but can know the fullness of energy experienced in living or growing with others.

If this other energy needs a name, then it could be called "the energy of an encounter." One does not encounter the entities or the materials one burns in order to extract energy from them; in the work of such destructive extraction, driving what it works upon to the physical end or to the point of depletion, there is not a smidgen of either interplay or fulfillment. An encounter happens only when nothing is missing, when I do not really need the other whom I encounter, and when I do not debate with myself what I can obtain from the encountered being. It happens as it did one foggy morning, when two people passed each other in silence in the pine forest that was much more than a "natural setting" for this passage. Because it does not belong to the strict economy of setting-to-work within the spiral of productive destruction, the energy of the encounter is essentially additional, supplementary, but also—and thanks to this—more crucial than what it supplements. It is a surplus that exceeds the economic framework, which is probably why philosophers have been prone to tying it to the uneconomic activity of art.

Admittedly, Aristotle glimpses this additional energy, persisting when all the goals are accomplished and residing in the act itself: "the act is an end and the being-at-work is the act, and since *energeia* is named after the *ergon* it also extends to the being-at-an-end or complete reality (*entelecheia*)" (*Metaphysics* 1050a, 21–23). The energy of the end is not destructive; it does not need to negate anything in the state of complete reality. Nor is it restless; it does not have any outstanding goals to be reached. It has almost nothing to do with the Western notion of energy, save for the ideal of accomplishment, to which it still seems to pertain. The extra step to be made here is to situate the calm and positive energy of the end in the midst of finite life itself and to enjoy it in the encounters, say, with and through plants. If this sounds outlandish to some readers, they should only recall that something of this energetic suspension in the middle of life has been a part of Buddhist philosophy and practice (where "complete reality" coincides with the openness of Śūnyatā, usually translated as "emptiness") for thousands of years.

In purely grammatical terms, the nonextractive approach to energy is the truest approximation to what the word *energized* conveys. The *passive voice* here connotes my receptivity to energy, which, as in the case of plants, is not procured by destroying the energizing entity. But it would be a mistake to perceive in this passivity a mere opposite of activity. In fact, being energized in an encounter is conducive to a more active comportment than the most feverish act of working, because, unlike my work, it will not be extinguished in the final product (it has none!). Still, one must understand *how* to receive the energy of the encounter, in which circumstances, and within what limits.

The question that immediately arises in this respect is what to do with the additional, supplemental energy when nothing is missing. This question is a little misguided, because it is no longer a matter of doing, or, if it is, then one of a doing uncoupled from work. The energy of an encounter, or of the end in the midst of life, is energy without *ergon*, emancipated from its heavy essential task of "setting-to-work." At the same time, it is not exactly play, either. Work and play are equally manipulative toward the other: the former—for a determinate instrumental purpose, the latter—for no apparent reason, for nothing. An encounter does not, by any means, preclude a playful indeterminacy, but it also cannot be slotted on the continuum of work and play, because its ethical dimension lies outside this continuum. I might as well refer to the effects it emits in terms of an ethics that does not burn (nor blinds) the energizing and the energized, in contrast to the most famous analogy of the ethical, the *ana-logon* that is the Platonic Sun.

Any description of alternative energy is doomed to be insufficient, if separated from the encounter wherein it germinates. And it might be especially astonishing to realize that this kind of energy does not obey the laws of its physical counterpart: the more it is spread, the more it is shared—the more it grows instead of diminishing or remaining constatnt. Dealing with it is not enacting, but engaging in the etymological sense of engagement as "giving a pledge," so as to spread and share this energy, which may not be detained or appropriated, least of all by only one of the participants in the encounter.

As for plants, they do not accumulate solar energy in their fruits and leaves, as economically inflected technoscientific jargon makes it appear. Instead, they preserve this energy of and for life and share it, exposing themselves to the world and releasing breathable air. Their engagement with the elements is part and parcel of the energizing encounter, the first encounter of the living with the other. In *Plant-Thinking* I wrote that "their unique ensouled existence enjoins plants to be the passages, the outlets, or the media for the other."[3] I must now specify that they are *energetic* channels, passages, outlets, or media, that are, simultaneously, ethical. Rather than a vegetal soul setting itself to work in the body of a plant, the life of this plant is a conduit of nondestructive energy to which other vegetal, animal, or human beings can also be privy.

This means, in turn, that our ethical categories need to be rethought, so that the habitually dismissed "pure means" (read: media or channels) would be more in tune with the ethics of life than the fictitious ends-in-themselves. Such "ends" are the blockages of energy, preventing it from being shared and spread. The emphasis on appropriation is the foremost example of blocking energetic channels and detaining what could be shared in the form of property. We stumble upon another paradox: if it is accumulated and kept, the energy of the encounter vanishes; if it is let go and passed on to the other, it thrives, together with life itself.

15–26 June 2014

13

COULD GESTURES AND WORDS SUBSTITUTE FOR THE ELEMENTS?

In an effort to understand reality, Western philosophy has recommended a method that is rarely doubted, even in the more critical strands of our intellectual tradition. To make the world palatable and comprehensible, thinkers, from the Pythagoreans and Plato to Schopenhauer and Nietzsche, have endeavored to translate it as a whole into symbols and codes.

More often than not, the scientific codification of reality resorted to numbers, totally indifferent to what they quantified. Sensible qualities dissolved in these empty universals that aimed to supplant, among other things, the elements, within which and thanks to which life unfolded. At times, the code was musical, transforming the world into a symphony. From the ancient "music of the spheres," conceived as an auditory expression of divine harmony, to Schopenhauer's transposition of musical scales and instruments onto a metaphysical hierarchy, with the basses representing crude materiality and the strings making audible the highest aspirations of spirit—from the one to the other extreme of Occidental history, the elements themselves underwent idealization as objects of hearing. All things considered, auditory translation is not so bad, especially when compared to the still more prevalent ocularcentrism permeating our metaphysics and privileging the most theoretical of senses (namely, vision) as a way of relating to exteriority. And it is not as atrocious as the scientific translation of reality into numbers, which did not leave anything whatsoever to our senses.

A definite choice is discernible in any substitution of symbols or codes for the elements. This choice expresses a preference for understanding the world over living in it, for turning it into an object of intellect over existing within its elemental milieu. Heidegger's hermeneutics, to be sure, insists, together with certain currents of Husserl's phenomenology, on the priority of a practical, lived interpretation *of what is*. Such an interpretation does not oppose itself to life, but supplements a particularly human kind of existence from within. As I have argued on several occasions, a lived hermeneutics is necessary to all living beings that must make sense of their milieu. Both plants and animals must continually interpret their surroundings in order to respond, adjust to, and thrive within constantly changing circumstances. Humans, in their turn, make sense of the world in words, forgetting, in the meantime, the nonverbal interpretations they share with all other living beings. As they search for a place of their own, they are expelled from the milieu of biological life and dismiss the varieties of sense that do not assume the shape of gestures or words as sheer nonsense.

Must every human act of meaning-making proceed on the terms that deny the inherent meaning of animals, plants, and the elements? Doesn't the incredible leaf of *Bryophyllum pinnatum*, also known as Goethe Plant, which is growing in a pot on my desk, hold an excess of meaning, to the extent that it condenses in itself the entire plant, with its growing, nourishing, reproductive, and other capacities? Is the Latin locution, *Bryophyllum pinnatum*, or any other name, including Goethe Plant, Air Plant, Life Plant, or Miracle Leaf, supposed to exhaust what this plant is? Are the study, measurement, and description of its biochemical processes adequate replacements for the experience of the plant I am contemplating now, let alone for its own mode of accessing the world?

Much more is at stake here than a mere critique of nominalism. Consider giving. When I have words to give, I usually destine them to other living human beings or domesticated animals, unless, poetically, I address myself to a mountain, a river, a forest, a flower, to the already dead and the still unborn, or, prayerfully, to God or the gods. But even for humans words alone are insufficient; their needs must be provided for before desire, in part verbally articulated, can express itself. What I give to plants under my care includes plentiful water, appropriate exposure to the sun, and the soil rich in minerals. That is to say, I give them a tiny portion of the elements, which they dispense back to me as a vegetal expression of light and warmth and water and the earth that helped them grow. As Thoreau put it in quasi-Plotinian language in *Walden* with regard to the beans he cultivated, his daily work was "making the yellow soil express its summer thought in bean leaves and blossoms . . . making the earth say beans instead of grass."[1]

To substitute words for elements, then, is to replace a human life for all life and, indeed, to favor only a small cross-section of human existence, while being inattentive to bodily needs and to our immersion in the elemental. This act of a global translation ineluctably impoverishes the sense of reality, at best calling whatever does not fall under its sway *inexpressible* or *mysterious*. By giving words as universal equivalents for everything and everyone that is, we take away from the world much more than we provide it with. Hence the effects of our actions are the mirror opposite of the behavior of plants that give back to the environment more than what they receive from it, making the elements themselves fecund and auspicious for the further growth of life.

In the case of gestures, the language in which they participate has more in common with vegetal expression than with words. Gestures are primarily spatial rather than temporal expressions, similar to the language of plants that articulate themselves in living forms, reiterated and replicated seemingly ad infinitum, as I argued in chapter 3 of

Plant-Thinking. In fact, gestures are more loyal to the sense of language as articulation than words: they articulate meaning by putting "this" and "that," or "this" *as* "that," together and they create jointures, *articuli*, that are as much spatial as they are semantic. Conversely, at the current stage of their use and abuse against life, to paraphrase Nietzsche, words disarticulate what they express, producing deeper and deeper fissures within and between the elements. They do not gather sense into a world that is or would be habitable for human and nonhuman beings alike, but contribute to the worldlessness of the elements, polluted and made sterile.

That the world is crumbling is not surprising, seeing that it has been analyzed by pure understanding hostile to life into inorganic chemical components. But, whereas plants recompose the minerals they nourish themselves on into new growth, the synthesis of understanding cannot give a livable character back to the world. Theoretical understanding grows, when it does, at the expense of life, which it breaks down into particles that are already dead and that have nothing to do with the elements. Through words and, above all, through numeric symbols, it substitutes life with death, which it then clothes in the dignity of a presumably higher existence, "the life of the mind." The classical correspondence theory of truth is absurd, precisely because it seeks to identify parallels between the sphere of the living, on the one hand, and a lifeless representation that becomes lethal upon contact with this sphere, on the other. While the earth says beans or grass or another plant depending on what grows on it, the truth of Western philosophers is tautological, saying nothing but itself.

Can I construct a universe in words—*the city of words*, as Socrates calls it in a different context—that will be so self-sufficient as to sever every one of its links to the elements? On the face of it, much intellectual energy in the West goes into such a construction, but, instead of expressing or representing what is, it interjects itself in place of what it was supposed to express or represent. The norms of an ideal world-construction demand that I nourish the others and myself on words alone, forgoing water, air, the warmth of fire, the earth, and the plants that enliven these elements, making them what they are. In response to this demand, I string words together, articulating them, and so formally imitating the spatial, material, fluid, and inherently meaningful articulations of existence or existences—mine, yours, of this dog, of this palm tree, and so forth. The replica will, nonetheless, be overshadowed by the power of understanding that supplants a livable universe, an animated *kosmos*, with a purely thinkable one and that idealizes relations among beings, slotting them into a homogeneous semantic network, in which they become tiny nodes or points of intersection.

If deconstruction manages to expose the instability of the code, to which reality has been reduced, it is because deconstructive readings shake up conventionally accepted semantic links and show that another node could stand in for any given presupposed unit of meaning. Derrida refrains from referring back to the elements or, for that matter, to anything falling in the scope of the outside-text, *hors-texte*, that does not exist. Rather, he reveals the play-aspect of understanding and intensifies this game to such an extent that its goal of making the world palatable and digestible becomes no longer accomplishable. The untranslatable and the incomprehensible *within comprehensibility* are posited as obstacles on the path of global translation, though they continue to participate, as negative modalities, in its logic.

By indulging in the excess of codifications, bordering on the indecipherable—in other words, by succumbing to a certain temptation of hypersymbolism—Derrida fatally wounds understanding detached from life, but, at the same time, falls short of questioning the primacy of the code or the cipher. With the view to helping deconstructive thought grow, I cannot evade such questioning that dares, at the extreme, to deconstruct deconstruction "itself." But a postdeconstructive break with conventional codification cannot be only theoretical; carrying it out means stepping outside, into a forest, a field, a garden, or a park. There it might be feasible to reconnect with the elements and with the vegetal world, and let the plants themselves express themselves, together with the earth in which they are rooted and which they "say."

How to put my words at the service of life, rather than charge them with the hopeless task of accurately reflecting it? I know what I *should not* do: attempt to drain the energy of plants and the elements into linguistic expression, which makes this energy deadly; accept codes and ciphers as something inevitable; fit the living universe within the chains of human signification; absorb the outside world into the rigid structures of meaning mandated by pure understanding and reproject these structures outward, anthropomorphized. I also have some preliminary indications concerning what I *should* do: go outside, expose my senses and thought to the elements and plants, let them express themselves in the nonverbal ways appropriate to each of them.

With this, I do not wish to preserve the dividing line between knowledge and mystery, rejecting the former in favor of the latter. Quite to the contrary, I want to affirm the lived interpretations and knowledges that pertain to nonhuman (elemental, vegetal, and other) worlds without either rendering them in words or relegating them to the incomprehensible. When I write that we ought to learn from plants, I am alluding to this affirmation of their wisdom, as well as to the task of finding a path toward it, that would not thoughtlessly follow the highway of a semantic translation. There is no secret recipe for imbibing the lessons of plants and the living energy of the elements, except that you must persevere as their apprentice without a term of maturation, just as you must keep practicing the art of living, without relinquishing the idea that you might forever be a novice in that art.

20 July–10 August 2014

14

FROM BEING ALONE IN NATURE
TO BEING TWO IN LOVE

Today most people both within academic and nonacademic circles confine a possible relation to nature to a false set of alternatives. Some believe that separation from the rest of the natural world is the destiny, if not the unchosen fate, of human beings (indeed, of the human *species*, as contradictory as that may sound). In strictly philosophical terms, such separation posits this world as an object over and against human subjects, who can then manipulate, mutilate, and ruin it as they please. While this view predicates progress on the exacerbation of our detachment from nature and "liberation" from its constraints, more and more the fate of humanity comes to resemble that of tragic heroes who unwittingly harbor the cause of their own downfall. Others advocate a complete immersion in the natural world, a return to the presumed simplicity of previous epochs in human history that were more integrated within the rhythms and cycles of the universe. This group is a minority, and its position is often ridiculed as a kind of neo-Luddite stance hostile to the cumulative achievements of the past. Their approach is more thoughtful, but insufficient, both because it represents a purely reactive response to the status quo and because it assumes that a ready-made solution to the current deplorable situation exists without the need to seek out a new path toward nature and toward ourselves.

Interestingly, it is the advocate of a return to nature who becomes the genuine tragic hero nowadays. Nowhere is this predicament more obvious than in Thoreau's *Walden*, which emphasizes the virtue of self-sufficiency in a temporary departure from "civilized life." "When I wrote the following pages," Thoreau begins his narrative, "I lived alone in the woods, a mile from any neighbor, in a house which I had built myself, on the shore of Walden Pond, in Concord, Massachusetts, and earned my living by the labor of my hands only."[1] He is entirely correct to suggest that, before living together with others, one must learn to live with oneself—hence, his predilection for the "solitary dwelling" in nature.[2] But because others have not yet learned to live with themselves, it is doubtful that one can ever cooperate with them in the art of living (together). Sadly, being alone in nature can repeat, albeit on a higher plane, the lonely hubris of the human confrontation with nature. Moreover, there is a fair chance that this would happen if the loneliness I embrace is a sheer reaction to the prevailing "species solipsism," of which it is a photographic negative.

Assuming that the choice between the arrogant opposition and the dream of a return to nature is a false one, what other options are plausible? Could it be that we are still unfamiliar with nature as a result of our excessive separation from or immersion in it? Too detached from the world around us, we are simultaneously too close to it, as we fail to touch, hear, or see this world, but only see, hear, or touch *through* it, as a tool or a medium for the accomplishment of our plans and desires. This is the dynamic of "dedistancing," *Entfernung*, that Heidegger outlined in *Being and Time*.[3] Similarly, when we are engrossed in nature, experiencing what Freud calls "an oceanic feeling" toward it, we fail to encounter it. Even if such engrossment were not complete, it would not lead to an encounter of or with nature, unless one reverted to what is to be encountered together with a human other.

I have already written earlier in this book, as well as in *The Philosopher's Plant*, that I am never really alone *in* nature, but alone *with* it.

In a different context altogether, Arendt was familiar with something of this distinction when she accentuated the contrast between loneliness and solitude. To attain a state of solitude, one must be already skilled at being with oneself (as Thoreau would also agree), whereas, in loneliness one is lost, in the first instance, to oneself. Solitude in nature, therefore, presupposes the maturity of living with oneself that, at the same time, invites coexistence with nonhuman others—the trees, the moss, the birds, and even the stinging nettle and the mosquitoes. Such living with . . . precedes every consideration of utility or harmfulness, which is why it equally affirms the entire natural world, without exception. And this is not to mention that it is from plants that I learn the tenets of living with . . . seeing that they gather the elements, live with themselves as with others, and coexist with others as with themselves.

So, rather than loneliness, solitude is my starting point in the midst of vegetal life. Next, however, I realize that I must start in solitude, but cannot continue (let alone accomplish anything, including the questioning of the very logic of accomplishment with its objective products and results) by myself. I reject the sham heroism of separation from nature and the tragic heroism of a nostalgic return to it as a fictional lost object. Nature has not been lost, because we have not found it yet. Everything is done either in preparation for finding it or in an effort to thwart this event. In solitude I discern the incompleteness of my experience (above all, of vegetal life), which must be repeated or reexperienced with another human being to make an encounter possible. An encounter nestled within an encounter: with plants and with another solitude: that of the other.

In an encounter of individuated human beings with each other and with nature, something remarkable happens: nature, too, becomes individuated, not in the sense that it is personified or anthropomorphized, but in the sense that it finally receives its due, neither more nor less: I can finally relate to this marigold as *this* marigold. A faithful approach to it is an issue of justice as much as it is one of a more complete experience. When I say that plants have their own kind of wisdom, I do not project some features of human intelligence onto the vegetal world. Instead, I acknowledge the environmental sagacity of the plants' relation to the elements, their contribution to the atmosphere, and their promotion of life. Respecting the uniqueness of the vegetal world is conducive, simultaneously, to an ethical approach to plants and to developing ontological or epistemological maturity, an eagerness to engage with each form of existence without deriving from it an abstract common denominator, which would be ultimately traceable to human values and representations. But why is a loving encounter with the other indispensable to the cultivation of such respect? Why can it not flourish in solitude?

Much depends on what is understood by the word *love*. In *Totality and Infinity* Levinas thinks of love as something ambiguous, "*the equivocal* par excellence," because it is "a relation with the Other that turns into need," going unto the Other, yet treating the beloved as an object to satisfy this need and therefore reverting back to the self.[4] That, in my view, is not love. Rather, love nurtures and safeguards the otherness of the human other, refusing to reduce her to the same. In love, which I like to think of as obscure warmth, two solitudes encounter one another, such that each one maintains the ability to live together with oneself. Love is a sharing of solitude(s); it is a misnomer to use this word to describe a state of fusion, including a fusion qualified as "ambiguous," encompassing the lover and the beloved. That is why love, in which two do not dissolve into one, is propitious for a cultivation of respect toward the natural world. Learning how not to reduce the human other to the same, I can hone the capacity not to impose such a reduction onto other living beings, be they animals or plants.

As I describe these passages—from solitude with nature, to the sharing of solitude with a human other, to a reexperiencing of nature through the shared solitude of love—I do not wish to suggest that a cause-effect logic is at work in each transition. It is in terms of a "virtuous" self-reinforcing circle that I consider the movement from the one in the midst of the plentitude of *phusis* to the two facing each other in love and on to the two facing each other and *phusis*. My solitude with nature intimates to me how to love another human better, while love, in which my experience is no longer partial (or simply *mine*), shows how to respect the uniqueness of each nature better. Far from being stages that need to be overcome, the three moments, between which transitions occur and passages open up, are the mutually complementary and equally vital aspects of being and becoming human.

In the history of Western philosophy, plants have been conceptualized as indifferent, insensate beings, wholly incompatible with a loving comportment. Avicenna is perhaps one of the few exceptions to this general rule, insofar as in his *Treatise on Love* he concedes that a vegetal "impulsion" is an expression of the plants' love, oriented toward water, sunlight, nutrients, and ultimately whatever is good for them. Admittedly, Avicenna does not succeed in decoupling love from need, even if its final objective is what lies beyond need, namely the good. Nevertheless, his *Treatise* indicates a certain proximity between love and vegetal being.[5] In what does this proximity consist?

Loving another human means, besides taking care to preserve her or his otherness, that I do not seize from the other more than what I give back and, indeed, that all such accounting becomes obsolete, so that I am ready to give incalculably more than what I receive. Here is an approximation of human love to the behavior of plants, which also dispense to the world much more than what they draw from it: they bestow upon it both breath and life. In another way vegetal love infinitely tends to its other, namely to sunlight, just as the lover strives toward the beloved without ever destroying the beloved's solitude. If human beings maintain themselves *as two* in love, they reiterate the capacity of the vegetal world (and of the plants' love, as Avicenna defines it) to be two with the world of the elements, to grow toward the elemental realm without transforming it into an object for appropriation.

28 August–8 September 2014

15

BECOMING HUMANS

All living beings participate in the process of becoming, which is, to some extent, synonymous with life. One aspect of the nihilism characteristic of Western philosophy was to freeze this process in the fixity of being, essence, or the categories, supposed to capture stable identities that exceeded and resisted the living flux of becoming. The ideally unchangeable remainder was taken as something that really was (is and will be) exempt from the order of time and earthly existence. But it is not sufficient to insist on pure becoming as the counterweight to pure being; even though they supposedly represent these two philosophical extremes, Heraclitus and Parmenides say, at bottom, one and the same thing, seeing that they make each of the terms secondary vis-à-vis *logos*. Now, at the close of the metaphysical era, Nietzsche's ingenuous solution to the problem was to consider the will-to-power in terms of "the being of becoming," as Heidegger elucidated in lectures on his predecessor's thought. Still, Nietzsche's suggestion is not so different from that of Hegel, who combined being and nothingness in a dialectical movement of becoming at the outset of his *Logic*.[1] And, if both attempts at a reconciliation fail, they do so because, as I have just noted, there is nothing to reconcile an abstract becoming that is but a mirror image of abstract being, considered under the auspices of another master concept, be it *logos* or Spirit.

Although Heidegger has been often faulted for a similarly abstract approach to the question of being (Levinas, for one, reproached him for constructing an ontological "totality"), his thought can point us in the direction of a more respectful attitude toward the incomparable ways of becoming, distinguishing various forms of life. In other words, it may allow us to register the uniqueness of human becoming, without, by the same stroke, denying the distinctiveness of animal or vegetal becomings, which are not at all at odds with the question of being.

First, a semantic difference. We speak of an animal becoming or a vegetal becoming, but, in our case, saying "becoming (a) human" is more apposite. The very idea of becoming-animal or becoming-plant, which Deleuze and Guattari tout in their *A Thousand Plateaus*,[2] makes sense solely to and for humans, eager to divest themselves of their identity as such or to occupy the place of other beings. Whereas a plant is a plant and an animal is an animal throughout their lifetimes (despite their capacity to acquire new experiences and to learn), for the human (for every one of us, singularly and together), humanity is a task to be achieved. I, you, she, he must work on it and expend energy that would go not into this or that external product we create, but into who or what we must become. Up to and including the creation of this very "we," in which the energy of each would be transformed into synergy and allow us to live and grow together. Heidegger's insight that a human is a being, for whom being itself is an issue or a question—a being who is a question or an enigma for itself—is highly relevant to this task.

Second, habitually, human becoming (better: "becoming (a) human") has been thought of as self-fashioning in opposition to the forces of nature—the raging elements, the tyranny of needs for nourishment and shelter, the threat posed by predators, and so forth. And, since plants are the foremost representatives of the natural world, our becoming has been connected to the inversion of their vegetal growth. Besides Plato's depiction of the human as a "heavenly plant," with roots suspended in the realm of ideas and branches growing down to earth (in *Timaeus* 90a), the Hindu tradition narrates about the tree *aśvattha* in Kaṭha Upaniṣad (6.1–4) and the Bhagavad Gītā (15.1–3). This eternal peepul tree similarly grows upside down in relation to the earthly vegetation: its roots are on high and the branches point downward. To perceive this vegetal representation of the cosmos, one needs to abandon the perspective of plant growth as it unfolds on earth, just as decisively as one is called upon, by Plato, to turn the order of vegetal life on its head in order to glimpse the realm of Ideas. The metaphysical version of becoming (a) human enjoins us to gain access to the esoteric essence of "true reality" by growing *against* the tendencies of plant growth. Nonetheless, the hold of plants on our senses and imagination is not diminished in all these negations and inversions: both the world and human beings, both thought and imagination, remain beholden to plants, albeit in a mutated and mutilated form.

Heidegger does not directly comment upon this predicament of thinking, even if he enables us to realize that it belongs in the long list of the metaphysical misconceptions of being (here extended to the East, too). The dynamics of becoming (a) human, in keeping with the metaphysical injunction, are self-exhausting: yearning for the security of being as immutable reality, wherein the mutated tree of the world or of humanity is rooted, becoming itself becomes otherwise than it is. Relegating truth to an atopic place outside, above, or beyond here and now, it divorces being from existence. Heidegger, on the contrary, proposes to locate the being of human beings in finite existence, in the limited terms of our lives. With this he dispenses back to us the possibility of growth and becoming that are not actualized either in a separate ontological realm or in the nothingness of death. Consequently, our open-ended growth within the finite horizons of our being inches closer to, without ever merging with, the becoming of plants.

Instead of growing *against* the vegetal world, which is a synecdoche of nature, becoming (a) human in the context of earthly existence means

use this for post

growing ~~with~~ plants ~~and all other living~~ beings. *Nota bene:* becoming with a tree, I do not ~~become a tree, but respect its potentialities and possibili-~~ ~~ties, which are different from my own.~~ Were I to conflate my becoming with that of a tree, I would merely be imposing my own measures and categories on this plant and losing myself in the process. Having said that, the difference between the tree and me is not unbridgeable, not because we share certain particles in our chemical composition but because, at the most basic level and in the first instance, I receive the gift of growth (and the very chance for becoming) from plants. It is from them that I must depart and to them that I must return on all the trajectories of my existence.

Upon a closer examination, growing and becoming with the other—a plant, an animal, a human . . . —is presupposed and included in every becoming-against. To oppose something or someone, I must already form a minimal community with it, him, or her; I must stand against a common horizon on a common plane and distance myself, commencing from that shared ground for negation. Becoming- or growing-with is, therefore, not a utopian design for the peaceful coexistence of all living beings; it is a description of what happens, even when murderous violence and oppositional attitudes prevail. This feeling of a minimal community, before its differentiation into positions *with* or *against*, need not be invented but only nurtured so as to let our becoming approximate the symbiotic logic of life.

What about the classical (Aristotelian) definition of the human as *zōon logon echon* in *Politics* (I, 1253a)? Normally translated as "an animal that possesses speech," this definition seems to be both biologicist and metaphysically inflected. But what if Aristotle names not the end product of human being but rather the infinite process of becoming (a) human, whereby the life, *zōe*, in us—in the "living thing," *zōon*, that we are—appropriates *logos* (speech, discoursing, thinking, or the voice . . .)? Clearly, such becoming is both necessarily incomplete and at least two-dimensional: while the living beings that we are claim *logos* for themselves, so, too, *logos* appropriates the life in us. The question is: How to proceed in light of this last appropriation? Does human becoming, or becoming (a) human, necessitate the supplanting of life with the *logos* that contravenes and negates our living energy, our "animal" vitality? Or, does it mean that, gradually, the inner articulation (*logos*) of a life deemed specifically human is revealed and, being thus disclosed, comes into its own (*echon*) in each one of us and in the human *qua* human? Indeed, the temporalities, the rhythms, paces, or tempi of becoming in the three processes—living, speaking-thinking-voicing, appropriating— are quite incommensurate with each other. Does becoming (a) human imply if not the harmonization then the ongoing finetuning of all these rhythms so that they might work in concert?

That things are not as simple as conventional translations of Aristotle have made it appear is evident based on the fact that *logos* is not an object or a thing to be turned into a possession. Even as a capacity, it requires lifelong development through a learning that is not equivalent to the formal process of education. Since it is not an object, coming to have, possess, or appropriate it will never culminate in absolute success, no matter how "skilled" one is in exercising *logos*. At best, coming-to-have without fully appropriating *logos* could be definitive of human becoming. But becoming (a) human is much more than that—the art of living with and between *zōe* and *logos*, privileging neither the one nor the other and striving to comprehend the one *as* the other. Needless to say, such an art of living cannot be practiced without the cultivation of sensitivity to the *logoi* of other living beings: animals or plants.

It is crucial to recognize that, in nonhuman living beings, life appropriates other kinds of *logoi* differently—certainly not by trying to transform them into pieces of property (as I have intimated, this approach is not entirely accurate for describing the human relation to *logos* either) or into inalienable possessions of what is most proper to them. Such recognition is invaluable for the process of becoming (a) human because it involves respect for the singular *logoi* of plants and animals. In the *Enneads*, Plotinus is exceptionally attentive to the silent *logos* of plants—which he also ascribes to the growth of nature as a whole—and their corresponding "growth-thought," *phutikē noesis* (III.8.8, 10–20). Human maturity, the sign that becoming (a) human is well under way, consists in resisting the urge to judge plants and animals by human standards and respecting the silent flourishing of plants and nature *as the manifestations of their becoming, without which we, too, cannot live.*

In a sense, plant life *is* their *logos,* and their becoming unfurls in this *is,* in this silent *logos* of life itself and in the finite life of *logos.* Only humans have erroneously replaced the mutual predication of life and *logos* with the relations of appropriation, possession, having. Heidegger was correct to note in *The Fundamental Concepts of Metaphysics* that animals (and plants) have the world in the mode of not-having,[3] but this is, far from being a defect plunging them into the state of world-poverty (*weltarm*), their strength. Humans should also strive to relate to the world nonpossessively; our becoming hinges on the cultivation of this approach in ourselves. If time were not so limited by the severe environmental crisis unleashed by humans, it would have taken centuries to correct the mistakes of metaphysics, to make our *logos* neither a murderous instrument, wielded against other living creatures and against ourselves, nor material for appropriation. But we do not have centuries to clean up our act. On a faster track toward a new becoming, plants can show the way: from them, we can begin to learn how to become (a) human within the context of our life that *is* our *logos*—the life/*logos* that is embodied, finite, necessarily shared . . .

15–30 September 2014

16

CULTIVATING AND SHARING
LIFE BETWEEN ALL

Throughout the encounter that is this book, one thought, one insight, keeps reemerging in various guises and formulations, be they experiential or more theoretical, namely that plants together with, and as, the elements of our world bestow the gift of life upon us. The life they gift us with is both of a physical and a spiritual variety. Through the nourishments to be digested and fresh air to be inhaled or exhaled, they give birth to the living spirit that is presumed to be intrinsically human.

With a careful elaboration and cultivation of receptivity toward, and active contemplation of, the vegetal world, it is possible to make a transition from "natural existence" to "spiritual life," taking care not to sacrifice the one to the other. Already Aristotle, in *De Anima*, noticed how the nutritive or vegetal soul—*tō threptikon*, which I have already invoked in this text—was responsible for enlivening, among others, the creatures that do not strictly belong to the botanical world. "The nutritive soul [*threptikē psukhē*] belongs to all other living creatures besides man, and is the first and most widely shared [*kai prōte kai koinotate*] faculty of the soul, in virtue of which they all have life" (415a23–26). All other modes of living, including the human, spring forth thanks to the sharing of plant vitality outside the confines of the vegetal world. To come back to this world, on a daily basis, is to revisit the generally unrecognized cradle of our own lives. To smell, taste, see, and perhaps touch plants (for instance, walking with our bare feet on the grass) is to be reunited with every bodily sense that vegetal sharing awakens in us. To think and dream with and of plants is to delve into a profound source of thinking and imagination. To be in touch with other humans through the vegetal world is to recover a distinct feeling for a community, no longer or not yet limited by city walls or virtual channels of communication.

In addition to our vitality, our very capacity to share may have been inherited from plants. The vegetal soul in Aristotle is not shared a posteriori, after some time has elapsed between its first affirmation in plants and its resurfacing in other beings. Sharing or being held in common (*koinotate*) is actually the essential feature of this soul, listed alongside its "firstness." Human beings, too, can live thanks to the energy we receive from the vegetal faculty of our souls and we can share life because we participate in the ontological self-giving of plants. I need not repeat that, becoming with and thanks to plants, humans do not become like them. Insofar as it is not recognized as such, the gift of vegetal life all the more approximates the nature of a gift outside the circuits of exchange. Further, it does not limit human beings to the figurations or processes of plants, but liberates us to be, or to become, ourselves. Plants give their gift without knowing it, whereas we receive it without being aware of this reception. That is how (and where) a living sharing of life begins.

It is inevitable that we circle back to the vegetal beginning in attempts to preserve, enrich, and share life. Each time the guiding question of these returns will be: How to receive the generous gift of plants, which we are incapable of containing fully, as we are incapable of containing life itself? This inquiry can even become our guiding star in everyday activities and in our highest (say, philosophical) pursuits alike. For instance: How to receive the overwhelming, overflowing gift in our dietary or even culinary practices? What to do with it, and what *it* can do with or to us, so as not to betray the giving of the vegetal gift?

With respect to the highest philosophical quests, Cicero, who was influenced by the Greeks in this and other matters, had an inkling that plant processes were fundamental to the education of the human soul. *Ergo,* his definition of philosophy as a *cultura animi,* the culturing or cultivation of the soul. In *Tusculan Disputations* he writes: "As a field, though fertile, cannot yield a harvest without cultivation, no more can the mind without learning; thus each is feeble without the other. But philosophy is the cultivation of the soul [*Cultura autem animi philosophia est*]. It draws out vices by the root, prepares the mind to receive seed, and commits to it, and, so to speak, sows in it what, when grown, may bear the most abundant fruit" (2, V, 13). There are quite a few problems with this enunciation of *cultura animi*: the idea that the soul is a passive substratum, which lies beneath as the ground for cultivation, or the fixation on fruit, prioritized over other vegetal phenomena. Yet, Cicero's expression is invaluable for the efforts to understand and to practice the cultivation and sharing of life.

Although, in the course of finishing *Plant-Thinking,* I did not have Cicero in mind, it is evident that his proposal fits exceptionally well with the final section of the book, titled "Philosophy, a Sublimated Plant-Thinking."[1] Philosophy as the cultivation of the soul has nothing to do with polemics, disputations, or debates struggling to capture an already established truth. Nor is it a merely propaedeutic, pedagogic discipline, though this view already veers closer to Cicero's. Philosophy, in Cicero's hands, becomes an aid for human becoming. It prepares the ground for the coming humanity, albeit not by separating us from our natural surroundings—and especially from the vegetal world—but by rediscovering something of vegetal life within the human soul (or soil) and by helping humanity make good on its promise, to flourish better in continuity and contiguity with plants. Ultimately, philosophy, the love of wisdom, cultivates the soul by showering light and heat onto it, that is to say, the very things that plants also require for their growth. I do not want to suggest that obscure warmth emanates exclusively from love and that brilliance radiates solely from wisdom. Both aspects of philosophy combine heat and light, the life-giving fire akin to that of the sun. Far from being relegated to the sphere of Ideas or some other "beyond," the positive solar philosophical endeavor belongs in the soil of the soul, which it patiently readies to receive the gift of plants, namely the seeds that will supplant vice. Is the vegetal derivation of Cicero's metaphor (and it is by no means certain that this is *just* a metaphor) responsible for bringing philosophy back down to earth, making sure, at the same time, that its fiery, solar side remains intact?

Cultivating life and sharing it between all might become a reality on the condition that a certain version of the ancient *cultura animi* would extend to all living beings. For this to happen, the vegetal emphasis already prevalent in Cicero ought to be intensified. Cultivation would not be internal to the soul, or, better still, the soul itself would not be restricted to a space of subjective interiority but would span the times and spaces *between* the living, whether they are plants, animals, or humans (call it the "intersoul"). Such a view of the soul already includes the principle of sharing, instead of appropriating life, often at the expense of other creatures. The cultivation of intersouls would mean nurturing the differences between the living, where the times and spaces that separate them retain their asynchronous and heterotopic specificities. In other words, as always, cultivation would not thrust its own arbitrary forms onto what or whom it cultivates, nor would it decide on what or who deserves to live and what or who should be uprooted, clearing the earth for the "monocultures" of vegetal crops or of human formations. Rather, it would permit all the intersouls it works with to proliferate in the best possible way, in keeping with their singular potentialities.

What is required, then, is a *cultura animi* for the world, through which the world might be finally shared, maintaining the multiplicity of worlds that constitute it: yours, mine, hers, his, theirs, the plants', the animals'... For Cicero, *cultura animi* is philosophy, which implies that the first (negative) task of thought, befitting the world's *cultura animi*, is to avoid detaching itself from life-giving fire and from the other elements, as it does in the process of constructing its parallel universe. The positive interpretation of this task enjoins philosophy to associate itself, for once, with the living and for the sake of the living: to experience, contemplate, learn from, and appreciate the elemental and vegetal worlds that inspire and animate life. Philosophy in the service of life—a dream Nietzsche cherished in a different way—would be precisely this *cultura animi*.

I noted, in chapter 3, that "sharing life and breath augments and enhances the sphere of the living, whereas dividing life into the so-called natural or human resources diminishes it." So when I suggest that we should intensify Cicero's original vegetal emphasis, I am putting plants on the side of sharing, and the core of our Western metaphysical tradition—on that of division. On the one hand, the cultivation and sharing of life, allied with vegetal existence, inspires an ecological attitude; on the other hand, the division and resourcification of life are conducive to the entrenchment of economic rationality. Adapting one or the other approach to life is much more than a matter of personal choice, aesthetic preference, or practical ethics. At stake in this divergence is the future of human and nonhuman existence. Besides the nonrenewable and highly polluting sources of energy, such as coal, oil, and natural gas, life itself may reach a point where it becomes nonrenewable under the weight of the divisions and economizations imposed upon it. Assuming our place as a part of the elemental community of sharing that preserves the solitude of each—the community of living-with and growing-with that does not result in a totalitarian closure of the common horizon—may be the last chance for saving life from the abuses it has tolerated (and cannot indefinitely keep tolerating) from human beings. If there is another kind of enlightenment, an awakening still capable of reversing the deadly trend, which is probably as old as humanity itself, and bringing us back to

the nearly destroyed world, from which we have unwisely detached ourselves as much as we could, such an awakening will be vegetal. And if we
still hope to attain vegetal enlightenment, we can do so only together—
each with herself or with himself, with each other, and with plants. For
the first time, then, we would move, through vegetal being, toward the
plants themselves, toward life, toward the world, toward ourselves, and
toward other human beings.

2–14 October 2014

EPILOGUE

Dear Luce,

I have a feeling that an ending opens onto a new beginning. In this autumnal gloom, when trees are shedding their leaves and the sky and water become virtually indistinguishable from one another, the time of harvest has arrived. My *Philosopher's Plant* is about to be published, with sincerest thanks to you for your feedback on the final chapter. The last chapters of our own book seem to be complete . . .

But the fruit is not at all the end; harboring seeds, it holds the promise of a new germination, of regeneration, another growth. I wonder *how* our shared work will continue, but I have no doubt *that* it will continue. What will that "other beginning" be? Or, better, how shall we approach it, having passed through—and, always on our way, continuing to pass through—vegetal being?

For Heidegger, the other beginning is *Ereignis* (usually rendered as "the event"), which silently resonates with, calls, and turns back to the first beginning, *phusis*. In the other beginning, as this event, the growing of nature becomes, for the first time, itself. How shall we imagine the arrival of that other beginning after the passage *Through Vegetal Being*?

Warmest regards,

Michael Marder
14 October 2014

* * *

Dear Luce,

To begin again, I think that it is important to emphasize that the thinking and living of nature as an event has barely begun. Even Heidegger was incapable of envisioning such a thing. Our book was one of the first steps in this direction, and plants have helped us in this in various ways. That is so because *phusis* as an event of growth is most discernible in plants, but not only in them. In fact, the happening of *phusis* also proceeds in us, as us, between us and all the other participants in its growing emergence. It is futile to think such an event outside our relations with other beings—relations that are only conceivable in terms of a language or languages. This does not mean that everything can be translated into a human grammar or expressed only by humans. We cannot forget the language of plants if we want to relate to them, nor those of other kind of existence. Nevertheless, in human discourses and relations, too, there needs to be a space of hospitality open to the possibility of communicating with nonhuman forms of life. I think that *Through Vegetal Being* has at least indicated how to prepare such a space.

The way in which this space is prepared is not without significance. Our exchanges proceeded slowly, through the mail, which gave ideas the time necessary for their growth and maturation. Moreover, they were not ideas detached from life, in the sense of an auto-bio-graphy, presented in the shape of abstract recollection. If the chapters themselves were written only very gradually, at least on my part, that is because they relied, at the same time, on memories, on the contemporaneous experiences of plants, and on the ongoing exchange of reflections and responses to our respective texts. In other words, it was imperative to cultivate an openness to the vegetal world without taking it, or anything in it, for granted. On the basis of this experiencing or reexperiencing of plant life, it was also important to remodel human relations—something that has begun, perhaps only barely, in our book—so that they would be more conducive to a sharing that would no longer be economic, but, rather, ecological.

As I look back on the path already traversed, I thus see two sources of inspiration that, in equal measure convoked my part of the text, namely the imperative to respond to my experiences with and of vegetal life and to respond to your experiences, including, above all, with and of the same kind of life. It also took time to understand what such double responding entailed. Quite quickly, it became evident that the responses would be deficient if they merely attempted to mirror the other, be it vegetal or human. Even mutual mirroring, which would have produced an infinite mimetic effect, would not have sufficed. A metaphysical way of thinking and acting would have reasserted itself in it all the more vehemently.

Another insufficient response would have been a direct answer to what is recounted in your part of the book. Here the danger would be one of losing the vegetal world and relapsing into a hermetically sealed universe of *logos,* styled as a dialogue, that remains deaf to the languages of plants. Therefore, the challenge has been responding obliquely and, in this indirection, keeping what cannot be said to the other. Does such obliqueness not maintain, as much as possible, what is untranslatable in language, including, in the first instance, in the language of plants? What will be made, by you and by the readers of this book, of the silences, the lacunae, the gaps—between the chapters, for instance—where more is happening than in the printed words themselves? What will the dates, specifying the time of the calendar for each chapter, say or fail to say? The singularity of these time periods can be occasionally related to the so-called content of the text, but, more often than not, it bears upon the double experience I have mentioned in ways that will remain as concealed as the experience of plant life itself.

It is my hope that our book would open alternative horizons for relating to the vegetal world, to another human being, and to other humans. At the same time, the obliqueness and singularity of what our work

strives to impart to the other make these horizons fragile and perceptible only with difficulty. I think that this could not have been in any other
way: a clear set of prescriptions, the absence of a two-part structure, and
a straightforward phenomenological description would have spoiled
the entire undertaking from the outset. Now only time will tell to what
extent an encounter "through vegetal being" has made, or will make,
other encounters (as well as this being itself) flourish.

With cautious hope for the future,

Michael Marder
1 November 2014

NOTES

PROLOGUE

1. Luce Irigaray, *Sharing the World* (New York: Continuum, 2008), p. 11.
2. Michael Marder, *Plant-Thinking: A Philosophy of Vegetal Life* (New York: Columbia University Press, 2013).
3. Michael Marder, "Vegetal Democracy: The Plant That Is Not One," in Artemy Magun, ed., *Politics of the One: On Unity and Multiplicity in Contemporary Thought* (New York: Bloomsbury, 2012), pp. 115–30.

1. SEEKING REFUGE IN THE VEGETAL WORLD

1. Michael Marder, *The Philosopher's Plant: An Intellectual Herbarium* (New York: Columbia University Press, 2014).
2. Martin Heidegger, "The Origin of the Work of Art," in *Basic Writings*, ed. David Farrell Krell, rev. ed. (New York: Harper and Row, 1993), pp. 139–212.

2. A CULTURE FORGETFUL OF LIFE

1. Michael Marder, "If Peas Can Talk, Should We Eat Them?" *New York Times*, 28 April 2012, http://opinionator.blogs.nytimes.com/2012/04/28/if-peas -can-talk-should-we-eat-them/ (accessed 30 December 2013).

2. "The first review on Amazon, for instance, goes so far as to claim that this book can only be understood as a brilliant satirical hoax, and that Marder himself is the Alan Sokal of the twenty-first century. And yet, those more attuned to the history and vocabulary of posthumanist thinking will recognize many valuable ideas here, sincerely presented." Dominic Pettman, "The Noble Cabbage: Michael Marder's 'Plant-Thinking,'" in the *Los Angeles Review of Books*, 28 July 2013, https://lareviewofbooks.org/review/the-noble-cabbage-michael-marders-plant-thinking/ (accessed January 10, 2014).

3. SHARING UNIVERSAL BREATHING

1. Emmanuel Levinas, *Otherwise Than Being, or Beyond Essence*, trans. A. Lingis (Pittsburgh: Duquesne University Press, 1998), p. 182.

4. THE GENERATIVE POTENTIAL OF THE ELEMENTS

1. Michael Marder, *Pyropolitics: When the World Is Ablaze* (London: Rowman and Littlefield, 2015), p. xii.
2. G. W. F. Hegel, *Philosophy of Nature: Encyclopedia of the Philosophical Sciences,* part 2, translated by A. V. Miller (Oxford: Oxford University Press, 2004).

5. LIVING AT THE RHYTHM OF THE SEASONS

1. Friedrich Nietzsche, *Untimely Meditations* (Cambridge: Cambridge University Press, 1997).
2. Friedrich Nietzsche, *Human, All Too Human: A Book for Free Spirits*, trans. R. J. Hollingdale (Cambridge: Cambridge University Press, 1986), p. 115.

6. A RECOVERY OF THE AMAZING DIVERSITY OF NATURAL PRESENCE

1. G. W. F. Hegel, *Phenomenology of Spirit,* trans. A. V. Miller (Oxford: Oxford University Press, 1979).
2. Georges Perec, *Species of Spaces and Other Pieces*, trans. John Sturrock (New York: Penguin, 2008); Gaston Bachelard, *The Poetics of Space*, trans. Maria Jolas (Boston: Beacon, 1994).

3. Johann Wolfgang Goethe, *The Metamorphosis of Plants* (Cambridge: MIT Press, 2009), p. 1.

4. Ibid., p. 3.

7. CULTIVATING OUR SENSORY PERCEPTIONS

1. It is true, however, that in harsher climates there is generally a greater sensitivity to the elemental world and to the weather—a constant and reliable topic of conversation in the Canadian winter.

8. FEELING NOSTALGIA FOR A HUMAN COMPANION

1. See the introduction to Michael Marder, *The Philosopher's Plant: An Intellectual Herbarium* (New York: Columbia University Press, 2014), p. xviii.

2. Antoine de Saint-Exupéry, *The Little Prince* (London: Wordsworth, 1995).

3. Novalis, *Henry von Ofterdingen: A Novel*, trans. Palmer Hilty (New York: Continuum, 1992).

4. Saint-Exupéry, *The Little Prince* , p. 93.

5. Ibid., p. 35.

6. Ibid., p. 71.

7. http://www.lifenews.com/2012/10/29/environmentalists-promote-legal-rights-for-plants-nature/.

8. http://blog.timesunion.com/animalrights/the-folly-of-plant-liberation/4277/.

9. RISKING TO GO BACK AMONG HUMANS

1. Friedrich Nietzsche, "Preface," in *On the Genealogy of Morality: A Polemic*, trans. Maudemarie Clark and Alan Swensen (Indianapolis: Hackett, 1998), p. 1.

2. Samuel Beckett, *Three Novels: Molloy, Malone Dies, The Unnameable* (New York: Grove, 1958), p. 359.

10. LOSING ONESELF AND ASKING NATURE
FOR HELP AGAIN

1. Claudia Baracchi, *Aristotle's Ethics as First Philosophy* (Cambridge:Cambridge University Press, 2008).
2. Michael Marder, "Vegetal Democracy: The Plant That Is Not One," in Artemy Magun, ed., *Politics of the One: On Unity and Multiplicity in Contemporary Thought* (New York: Bloomsbury, 2012), p. 126.

11. ENCOUNTERING ANOTHER HUMAN
IN THE WOODS

1. Henry Thoreau, *Walden*, ed. Jeffrey S. Cramer (New Haven: Yale University Press, 2006), p. 289.
2. Francis Ponge, "From the Pine-Wood Notebook," trans. Derek Mahon, *Metre: A Magazine of International Poetry* 1 (Autumn 1996): 22.
3. Jean-Jacques Rousseau, *Reveries of the Solitary Walker*, trans. Russell Goulbourne (Oxford: Oxford University Press, 2011), p. 82.
4. René Descartes, *Discourse on Method and Meditations on First Philosophy*, 4th ed., trans. Donald A. Cress (Indianapolis: Hackett, 1998), p. 68.

12. WONDERING HOW TO CULTIVATE
OUR LIVING ENERGY

1. Cf. Michael Marder, *Pyropolitics: When the World Is Ablaze* (London: Rowman and Littlefield, 2015), p. 94.
2. "All variety is reducible to the notion of that which is *combusted*; some are conceived in reduction—(the phenomenon of this reduction is vegetation; at the lowest stage the vegetation of metals which are maintained by the inner glow of the Earth, at a higher stage the vegetation of plants)—others in permanent combustion (the phenomenon of this permanent process of combustion is animal life)." F. W. J. Schelling, *First Outline of a System of the Philosophy of Nature*, trans. Keith Peterson (Albany: SUNY Press, 2004), p. 96.
3. Michael Marder, *Plant-Thinking: A Philosophy of Vegetal Life* (New York: Columbia University Press, 2013), p. 42.

13. COULD GESTURES AND WORDS SUBSTITUTE
FOR THE ELEMENTS?

1. Henry Thoreau, *Walden*, ed. Jeffrey S. Cramer (New Haven: Yale University Press, 2006), p. 170.

14. FROM BEING ALONE IN NATURE
TO BEING TWO IN LOVE

1. Henry Thoreau, *Walden*, ed. Jeffrey S. Cramer (New Haven: Yale University Press, 2006), p. 1.
2. Ibid., p. 76.
3. Martin Heidegger, *Being and Time*, trans. John Macquarrie and Edward Robinson (New York: Harper and Row, 1962).
4. Emmanuel Levinas, *Totality and Infinity: An Essay On Exteriority* (Dordrecht: Kluwer, 1991), pp. 254–55.
5. Avicenna, *A Treatise on Love by ibn Sīnā*, trans. Emil Fackenheim, *Medieval Studies* 7 (1945): 208–28.

15. BECOMING HUMANS

1. G. W. F. Hegel, *Logic: Encyclopedia of the Philosophical Sciences*, part 2, trans. William Wallace (Oxford: Oxford University Press, 1975).
2. Gilles Deleuze and Felix Guattari, *A Thousand Plateaus: Capitalism and Schizophrenia*, trans. Brian Massumi (Minneapolis: University of Minnesota Press, 1987).
3. Martin Heidegger, *The Fundamental Concepts of Metaphysics: World, Finitude, Solitude*, trans. William McNeill and Nicholas Walker (Bloomington: Indiana University Press, 2008).

16. CULTIVATING AND SHARING
LIFE BETWEEN ALL

1. Michael Marder, *Plant-Thinking: A Philosophy of Vegetal Life* (New York: Columbia University Press, 2013), pp. 170ff.

INDEX

Lightning Source UK Ltd.
Milton Keynes UK
UKHW021813050822
406910UK00009B/1230